Potential Energy

Potential Energy

An Analysis of World Energy Technology

MICHAEL KENWARD
Technology Editor, New Scientist

CAMBRIDGE UNIVERSITY PRESS

CAMBRIDGE

LONDON · NEW YORK · MELBOURNE

Published by the Syndics of the Cambridge University Press
The Pitt Building, Trumpington Street, Cambridge CB2 1RP
Bentley House, 200 Euston Road, London NW1 2DB
32 East 57th Street, New York, NY 10022, USA
296 Beaconsfield Parade, Middle Park, Melbourne 3206, Australia

© Cambridge University Press 1976

First published 1976

Printed in Great Britain at the
University Printing House, Cambridge
(Euan Phillips, University Printer)

Library of Congress Cataloguing in Publication Data

Kenward, Michael, 1945–
  Potential energy.
  1. Power resources. 2. Power (Mechanics)
I. Title.
TJ163.2.K46   621   75–36174
ISBN 0 521 21086 0 hard covers
ISBN 0 521 29056 2 paperback

Contents

[v]

Preface

The world's energy supplies come from the least plentiful energy resources. In the United States, for example, over 75 per cent of the nation's energy is consumed as oil and natural gas – both resources are dwindling. A similar state of affairs exists in most of the world. There are local anomalies – Britain has its North Sea oil, for example – but the long-term outlook is pretty much the same the world over. This situation will not change overnight. The key to future energy supplies is new energy technology. As technology becomes more complex and the energy system grows it takes longer for any technical innovation to have any significant impact on the system.

The upheavals that pervaded the world's energy markets in 1973 and 1974 brought home to everybody the importance of changing the direction in which the energy supply system is moving. There is no longer any visible shortage of oil or energy, so people assume that the 'crisis' is over and they can go back to their oil- and gas-burning ways. The artificial political crisis may be a thing of the past but the long-term problems are as severe as ever. We still have to find ways of using our abundant energy resources as efficiently as possible.

There are many ideas that could be pursued to help solve our problems – if we had unlimited time and money. We don't, so the problem of creating a sensible 'portfolio' of energy research and development (R&D) is pressing. R&D possibilities have to be assessed before we start throwing money into the world's energy research laboratories. The scientists and engineers who want money for their grandiose projects will not point out the snags looming on the horizon; indeed, they may not see them. I have tried to assess some of the better known possibilities. I have not set out to provide a catalogue of energy technologies, let alone a blueprint for an energy R&D policy. Instead, I have tried to point to the questions we must ask before we can be sure that energy R & D is anywhere near the right track.

The first questions are: 'What energy do we need? How will we use it?' The answers to these questions must shape any R & D programme. It is then important to assess the prospects for the various R & D options. Without this appraisal we can be fooled into supporting the most way-out and improbable options because they offer fantastic opportunities. If the chances of success are small, it would be unwise to rely too heavily on a potentially immense energy supply system.

We also have to reckon with the established energy system. If someone came along tomorrow with *the answer* to our energy problems it would be years before the new technology could have an effect on our energy consumption. And such a final solution could throw the existing energy suppliers into panic as they tried to earn as much as possible in the limited time left to them.

There is no shortage of questions to ask before we can begin to be confident that our energy R & D is on the right lines. We should, of course, keep on asking these questions. The R & D programme that is appropriate today may be all wrong in just five years time. Who knows what the search for new energy resources will throw up? R & D programmes must be modified in the light of their successes and failures, and the successes and failures of other projects, including the search for new energy reserves.

It takes time to develop new energy technologies. Politicians – the people who allocate R & D funds on our behalf – rarely look beyond the next election. It will be all too easy for them to cut off the flow of money when the energy crisis has receded even further into the past and out of the limelight. This must not happen if the world is to have the energy technologies it needs to take it into the next century. There are clear signs that the environment has lost much of its glamour as a popular issue. Just as the world must not resume its polluting ways, it must also maintain an active energy R & D programme that responds to changes in the world's energy system.

The numbers game —
some units and equivalents

Energy comes in a variety of different units – from the ton of coal and the barrel of oil to the more exact units of the physicists. Each industry uses its own units, and where possible I stick to those appropriate to the business I am describing – there wouldn't be much point converting everything into joules.

Let us begin with some standard numbers: I use the American convention where a thousand million is called a billion; a million million is called a trillion (a number they use most in the gas industry). I also use the prefixes mega- = million (10^6), giga- = billion (10^9), tera- = trillion (10^{12}). So a million tons can be called a megaton.

Electricity and its energy content can cause counting problems. Power station outputs are usually given in watts – megawatts to be more precise – but this is actually a measure of power, not energy. Power is turned into energy when it is applied for a period of time – so a unit of energy is the watt hour (Wh). We talk about power stations producing so many megawatt hours (MWh) of electrical energy.

Other energy systems use watts and watt hours, so when there could be confusion you will see power described as watts (thermal) or watts (electrical) – these can be written as Wth and We.

Because each energy system is described in terms of its own system of units, it might help to set down some approximate equivalents.

1 GWh of electrical energy is equivalent to:

135 tons of coal

80 tons of petroleum

3 million cubic feet of natural gas

This should not be confused with the amount of fuel needed to generate 1 GWh of electricity – the above is a measure of the energy contents of the fuels. For every GWh of electricity generated in a power station, something like 2 GWh of thermal

(heat) energy are wasted. So the amount of fuel needed to produce
1 GWh of electricity is roughly:

500 tons of coal

270 tons of petroleum

10 million cubic feet of natural gas

Other numbers that might be useful include the number of barrels
of oil in a ton of oil. This depends upon the sort of 'ton' you are
talking about. A barrel of oil is equivalent to:

0.136 metric tons (a metric ton – a tonne – is 1000 kg)

0.134 long tons (a long ton is 2240 pounds)

0.150 short tons (a short ton is 2000 pounds)

Coal is a far more variable quantity than oil – which itself varies
depending upon the source of the oil – and it is difficult to give
an exact comparison between the two fuels. Britain's Department
of Energy uses a conversion factor of 1.7 tons of coal = 1 ton of
oil. I use this conversion factor where appropriate, unless there is
a warning to the contrary.

Chapter 1

The energy picture

Energy crises are not new. We have had them before; but in the past they all eventually evaporated despite popular belief that each was, when it first reared its disturbing head, the energy crisis to end all energy crises. Each was completely unlike anything that had happened before; it was, at last, *the real thing*. And yet each crisis eventually fizzled out. What are the chances of the difficulties of the 1970s fading away as the political difficulties in the Middle East become history? Will the crisis mentality continue, or will this energy emergency go the way of its predecessors as the prophecies of doom and gloom come to nothing? Or will today's concern continue and manifest itself in a new spate of energy research and development?

Much of the evidence adduced to confirm 'crisis seventies' as the true watershed is demonstrably false. There are, however, some signs that today's is a truly unique situation. The supply of energy is nowhere near being overtaken by demand; but that is not the sole indicator of problems ahead. Indeed, if we ever approach a state of affairs where we need more energy than the world can supply it will be too late to worry. One indicator that we have come to a crossroads, in thinking if nothing else, is the sudden awareness of the implications of sustained exponential growth. Another and more serious sign that we now face a new situation is the change in the rate at which new energy reserves are being discovered. In particular, the experience of the United States in oil and gas exploration tells us that the life of the world's oil reserves – the most popular fuel with the most rapidly growing consumption – could be very short, in modern industrial and financial terms as well as geological terms.

The change in the pace of oil discovery is significant because from now on, unless the rate at which new oil reserves are found takes a dramatic upturn, we shall be draining oil from a shrinking

[1]

store. New oil fields will be discovered, but unless they more than match consumption the net effect will be a continuously shrinking stockpile. This means that we are headed down the path toward an oil-less society. (Just when the oil will run out, and why this is a relatively new situation will become clearer when we look at the world's energy reserves in the next chapter.) But this is not, in itself, the catastrophe that the 'doomsters' would have us believe. Conventional oil reserves can be backed up by oil shale and tar sands to give us temporary relief; and there is no reason why we should not learn to live with a smaller supply of oil. It is true that the later we leave it before changing tack the harder we will find it to adapt to a new energy picture; but, by making a positive decision to opt for another path, we can minimise the changes that society will have to go through on its way to an oil-less existence. The size of the problem means that we cannot afford to squander the available resources by pursuing every research option and trying to do too many things at once.

The fuel mix

Before we can embark upon an extensive reorientation process for the world's energy system, we must know just what today's energy system is, what fuels go where, and how they are used. We must know what energy reserves are available and, perhaps more important, at what price they are available and with what technologies they will be exploited.

The first thing to look at is how the world's energy market is shared between different fuels, and how these fuels are used. Clearly the two are related – the use to which energy is put is related to the form of the energy. Coal-fired cars are a possibility, but with petrol refined from crude oil available at a reasonable price, there is little incentive to develop something messy and inconvenient like a coal-fired car engine. (Clearly one way out of this difficulty is to turn the coal into liquid fuel.) By the same token, despite a massive and expensive research and development programme the United States has found that nuclear-powered aircraft make little sense. However, for many energy users the switch from one fuel to another can be made easily enough. The gradual electrification of the energy system is one option that could shift the demand from fossil fuels to nuclear power. At the same time shipping could turn to nuclear power, and with the price of oil at its present level the nuclear ship is becoming an increasingly attractive proposition.

Knowledge of where energy is used is important for an evaluation of society's ability to switch from one energy system to another. It may be very easy to substitute one fuel for another for certain uses, but if the amount of energy involved is insignificant the substitution will do little to alleviate the world's energy problems. Thus it is an expensive luxury to initiate large research and development programmes that can have little impact on the overall energy picture. ('Research and development' is usually abbreviated to R & D for convenience.)

The world's energy system is not an easily observed affair. Statistics for the separate fuels are not always as reliable as they could be, and it is even harder to find out where fuels go once they have entered the system and are out of the hands of the fuel industries. After the upheavals of 1973/74, Britain, for one, found that it just did not know where industry used energy. Certainly each company knew how much it was spending on oil, gas, and electricity; but after the energy had arrived on the site few companies had much idea of where it was used. To begin with we need to know how much of each fuel is produced and consumed. The world relies on the primary fuels, coal, oil, and gas. Nuclear power may be the most publicised and researched of today's energy systems, but it has yet to make a significant impact on the world's energy consumption.

The United States dominates the world's energy economy. This may cause Americans some soul-searching, but it does make it reasonably easy to arrive at fairly reliable figures for the world's energy consumption. The US has a well-documented energy economy, as have most of the large energy consuming areas. The errors creep into the statistics in the smaller countries where consumption is low and there may be significant consumption of wood and equally unquantifiable fuels. It has become something of a cliché to retail the horrifying figure: the US has but a small fraction of the world's population (about 6 per cent), yet it devours something like a third of the world's annual energy production. The rest of the 'developed' nations together consume about the same amount of energy as the US. Clearly we can obtain a reasonably accurate view of the world scene, given that our picture of the situation in the developed world (the nations in the Organization for Economic Co-operation and Development) is acceptably accurate.

One difficulty soon crops up when you try to analyse energy statistics: energy comes in a variety of units, some of them of theoretical origin (such as the joule) others more practical (such as the barrel of oil). We even have to cope with a variety of tons – the metric tonne (1000 kilogrammes), the long ton (2240 lb), and the short ton (2000 lb). When trying to compare one fuel with another, a popular comparison unit is the coal equivalent tonne. Unfortunately different people attach different 'coal equivalents' to a tonne of oil. In Britain a tonne of petroleum is counted as 1.7 tonnes of coal; the United Nations uses a conversion factor of 1.3. Tables 1.1 and 1.2 of world energy consumption are taken from UN statistics. They show the impact of growth on energy consumption, as well as the dominant position of the developed nations. The units used in the tables give the energy as coal equivalents, in either millions of tonnes or kilogrammes.

Coal was the first fossil fuel to be widely exploited. It owed its development to a previous 'energy crisis' brought about by growing demand and dwindling supply – in the consumption of wood. When there was not enough wood to meet both the fuel needs of Britain's iron industry and other wood-burners, and when builders became competitors for what wood was left, there was a great incentive to develop coal. Thus the southern forests gave way to the northern coal mines as the source of industrial fuel, and England's industrial centre moved from one part of the country to another. We still find that despite society's greater mobility coal consumers prefer to go to where the coal is rather than to shift the coal. With today's large power stations, this trend is even more marked. If coal conversion technology is developed to turn coal into liquid and gaseous fuels, the conversion plants will almost certainly follow the South African pattern, where the coal liquefaction plants are built near to coal mines. We should not forget this when the idea of a new era of coal is advocated.

The world fuel statistics (see table 1.2) show the dramatic

Table 1.1. *Growth in world energy consumption over the decade 1962–1972*

	1962			1972		
	Total (mtce[a])	(%)	Per capita (kgce[b])	Total (mtce)	(%)	Per capita (kgce)
Developed countries	2782	63.2	4166	4604	62.1	6211
Developing countries	339	7.7	239	657	8.9	362
Planned economies	1281	29.1	1269	2149	29.0	1825
World total	4402		1423	7410		1984

[a] mtce = million tonnes coal equivalent.
[b] kgce = kilogrammes coal equivalent.
SOURCE: *World energy supplies 1969–1972*, UN Statistical Papers, Series J17.

Table 1.2. *World energy consumption, by fuels: over the past decade oil has overtaken coal as the most popular fuel*

	1962		1972	
	(mtce)	(%)	(mtce)	(%)
Solid fuels	2075	47.1	2407	32.5
Liquid fuels	1502	34.1	3220	43.5
Natural gas	728	16.5	1603	21.6
Hydroelectricity and nuclear electricity	97	2.2	180	2.4
World total	4402		7410	

SOURCE: as for table 1.1.

changes that have come about in the past decade or so, let alone over longer time-scales. The changes in the 1960s really were dramatic. To begin with the consumption of liquid fuels doubled in just ten years. Solid fuels maintained their production level, with a slight rise at the end of the decade. The share of the energy market taken by solid fuels fell from nearly half to slightly more than a third of the total energy consumption. Oil might have taken an even bigger share of the market had there not been the financial and social capital tied up in the solid fuel industries. This is certainly true for Britain where governments have been reluctant to adopt a strictly commercial approach to running the coal industry, and have kept the industry alive for social reasons as well as financial and energy reasons.

The fuel 'mix' at any time is a reflection of the price structure and its recent changes. One fuel may be cheaper than others at any instant, but the momentum of the system makes it impossible to switch from fuel to fuel at a flick of a switch. The statistics show that while the 1960s were the years of the atom as far as R&D was concerned, nuclear power accounted for a tiny fraction of the world's energy supply. Nuclear power hogged most of the world's energy R&D budgets, but the inertia of the electricity generating system was so great that only now is the atom making any inroads into the electricity supply industries of the developed nations. This underlines the fact that it really does take decades for a new energy system to make its presence felt where it matters – in the energy statistics.

We can hardly accuse the nuclear industry of being slow off the mark. Despite the difficulties of starting from scratch with a new technology, the industry has made significant progress. A closer look at the UN's energy statistics shows that in 1961 nuclear power stations generated 4358 million kilowatt hours (0.18 per cent of the world's 2453 thousand million kWh). In 1972 nuclear power stations generated 142132 million kWh. While this was just 2.5 per cent of the world's electricity generation of 5629 thousand million kWh, it represents a 32-fold increase in the generation of electricity by nuclear power stations. Thus a truly dramatic step forward in a new energy technology made an extremely small impact on the world's energy consumption.

Any new energy technology that requires extensive research and development and sizeable capital investment will inevitably follow a similarly protracted growth process. And as the existing system grows it becomes even harder for new technology to dent the statistics. To begin with it takes time to build up the industrial structure that brings about technological change. It took time for nuclear power to become competitive with fossil-fuelled power stations. Even before the oil crisis nuclear power stations produced cheaper electricity than oil-fired plants. Since the price of oil received its rude bump upwards, electricity utilities have been queueing up at the doors of the nuclear reactor makers. At the same time, national governments have expressed their desire to switch to nuclear power, and in some cases they want to order nothing but nuclear power stations in the foreseeable future. But few governments expect to be able to reduce their oil consumption in the near future. All they can hope for is a slower growth rate. In fact, most countries anticipate slower growth in all sectors of the energy market including electricity, which is taking some of the edge off the growth of nuclear power. Another restraint is the shortage of money for something as expensive as a nuclear power plant, which costs more to build than a fossil-fuelled power station, but which produces cheaper electricity because fuel costs are far lower.

The bald statistics about the fuel mix and how it is changing are instructive enough, bringing home as they do the inherent inflexibility of the system; but they are of little use when we want more detailed clues as to the possible future shape of the world's energy system. The old idea that future growth rates would not be too far removed from those of the immediate past has taken a sharp knock thanks to recent events. Clearly the future fuel mix will depend to a large degree on the reserves situation, which we will look at in the next chapter. The future will also depend on the uses to which fuels are put. Thus information on the consumption of oil, coal, gas, and so on, has to be backed up with information on the way in which these fuels are used.

International implications

The rich nations, with their massive energy consumption, will set the pace when it comes to developing new energy technologies. Clearly the dominant nation in the clique of well-off countries is the US, where the energy R&D bandwagon has gained the greatest momentum. This makes it inevitable that the future shape of the world's energy system is being shaped, to a considerable extent, in the US. But energy is an international business, and even the US acknowledges the need for coordinating national energy policies.

The international nature of the world energy system is well enough understood: it was, after all, the sudden halt in international oil trade that brought a growing unease about energy up to crisis status. In reality, the international nature of the energy industry is almost solely a result of oil movements. Coal is still a significant fuel, but it is mostly consumed in the countries where it is mined. Just 8 per cent of the 2500 million tonnes of coal mined each year is traded in international markets. This represents 7 per cent of the world trade in primary energy resources. A little over half of the world's crude oil production is traded internationally, and nearly 60 per cent of this international oil trade is centred on Middle East production. Thus international oil politics are at the heart of international energy policy.

The calls for an internationally agreed oil supply policy have been voiced often enough, but the solidarity evaporates when it comes to doing something positive. Indeed, during the 1973/74 disturbances governments were trying to steal a march on their neighbours. The oil companies had to step in and act as a sort of 'world government' for oil, shunting supplies around so that countries subjected to Arab boycotts obtained oil from countries that were not a party to the blacklisting. Thus Holland was hit less dramatically than might have been the case had the wishes of other oil importing countries, which wanted to get their hands on every drop of oil available to them, been respected. Even Holland's EEC partners showed little sympathy. Since the oil crisis there have been several attempts to establish international agreements on what to do in the event of future disruptions in oil supplies.

International cooperation has been more successful on the R & D front. The US and the UK have come together on coal research because the US has suddenly rediscovered the value of its huge coal resources. The US now wants to make up for years of scientific and technological neglect of coal. The UK may not have built large pilot plants to try out coal processing technologies on a large scale, but it has maintained a high quality programme of research at the laboratory level. By teaming up, the US and UK can help each other out.

Most countries now realise that they cannot afford to develop all new ideas that come from their research laboratories. Thus, no matter how popular energy R & D becomes, some projects will not be supported. This new reality makes it harder, though not impossible, for researchers to find funds for work that has no obvious potential. Energy R & D will be tailored to dovetail with a country's existing and anticipated energy pictures. A key factor in formulating an energy R & D policy must be an understanding of the ways in which energy is consumed.

The switch to electricity

The first significant trend that becomes obvious from a look at the fate of fuels is the growth of electricity in the energy systems of the developed nations. Energy consumption by the electricity supply industry has grown over the years. In the UK, fuel consumption by the electricity supply industry rose from 24 per cent of the nation's total energy supply in 1960 to 33 per cent in recent years. In the US more than 40 per cent of the nation's energy supply goes to electricity generation. Underlying this growth in the electricity industry's fuel consumption is an increase in the efficiency of generation, hence the growth in fuel consumption at the power stations is more than matched by the growth in power sent out from these stations. Between 1963 and 1973 the total fuel input to the electricity supply system in the UK rose by 45 per cent; electricity output rose by 65 per cent during the same period.

Clearly, new electricity generation technology could have a dramatic impact on the world's energy system. At the same time, we can expect a large share of the funds available for energy R & D to go to the electricity generation industry.

The whole question of the electrification of the world's energy system is a matter of intense debate. More conventional energy thinkers advocate a dramatic increase in electrification. Their argument is that, with a large electricity generating and distributing network, an 'electrified' country is in a position to ride out the fluctuations in the fortunes of individual fuels. Thus, as oil runs out electricity generation will switch to nuclear power. And when fusion reactors are introduced into the electricity supply industry fuel problems will become a thing of the past. The 'electric' strategy has been advocated by, among others, the Commission of the European Economic Community as a way of reducing the EEC's reliance on imported fuels.

Table 1.3. *Energy consumption by*
consuming sectors in the US

	1960		1968	
	($\times 10^9$ Btu)	(%)	($\times 10^9$ Btu)	(%)
Residential	7 968	18.6	11 616	19.2
Commercial	5 742	13.3	8 766	14.4
Industrial	18 340	42.7	24 960	41.2
Transportation	11 014	25.5	15 184	25.2
Total	43 064		60 526	

SOURCE: Stanford Research Institute.

Less conventional thinking is that electricity generation is an inefficient process, with two-thirds of the energy delivered to a power station wasted as hot air and warm water. This wasted energy should, the argument goes, be put to use, for heating homes, for example. That the cost of utilising this energy might be prohibitive does not come into the argument – it is energy that must be saved, say the extremists, not money. The advocates of cost-effectiveness do not approve of this uncompromising approach. To them electrification of our energy system could eliminate several painful steps along the path to a fission or fusion economy. For example, if we want to continue to drive around in cars when the crude oil has run out, we can either turn coal into oil, make hydrogen from water, or build battery-driven vehicles. The first of these options is fine as long as coal lasts, but even this will run out one day despite its large reserves. So we may be forced to turn to one of the other options in the end. If this is the case, why not go the whole hog now? Here we are entering the realm of energy consumption as well as production.

Tables 1.3 and 1.4 show where energy is consumed in the US and UK. These statistics show that dramatic fuel shifts can be achieved by changes in consumption technology as well as production technology. From these tables we can see that changes in transport technology, for example, can have a significant effect on the world's energy system. Looking at the UK statistics for energy consumed by transportation, it is obvious that petroleum is by

Table 1.4. *UK energy consumption (million therms)*
broken down into fuels and consuming sectors

	Coal	Oil	Elec-tricity	Gas	Total	(%)
Industry	6556	11390	2727	5109	25782	42
Domestic	5328	1668	3114	4815	14925	24
Transport and other	997	16565	1651	1083	20296	33
Total	12881	29623	7492	11007	61003	

SOURCE: *Digest of United Kingdom Energy Statistics 1974*, Department of Energy, HMSO.

far the most important source of energy. So a shift away from oil could be achieved by developing transport technologies that do not consume oil. Oil is the most endangered fuel as far as reserves are concerned, hence it makes a lot of sense to devote a significant effort to the problem of transport based on other energy sources. The 'electric' lobby has made its bid for this market by setting up extensive R&D projects on electric cars.

Table 1.5 shows how different industries eat into Britain's energy supply. Here too the numbers are a guide to where it makes sense to spend money on R&D in pursuit of new technology that consumes energy in a different form or uses less of it. On its own this table does not contain enough information to determine research priorities. The separate sectors do not contribute to the gross domestic product in the same way that they consume energy, and some industries are more adaptable than others. The food industry, for example, consumes half as much energy as the engineering industry, but it is not as amenable to technological changes that could change its energy pattern. Some industries are more energy intensive than others, and more easily influenced. The food industry is very fragmented so it is less easily changed than industries based on huge processing plants. The cement industry, for example, is a major consumer of energy, because of the energy intensiveness of the processes involved. Something like 25 to 30 per cent of the price of cement delivered in 1973 by the UK industry went to pay for the energy consumed in producing the

Table 1.5. *Fuel consumption in the UK (million therms)*

Sector	Solid fuel	Gas	Petro- leum	Elec- tricity	Total	% by sector
Agriculture	37	—	710	135	882	1.4
Iron and steel	3181	862	2123	362	6528	10.7
Food	388	272	1238	206	2104	3.4
Engineering	605	1077	1771	782	4235	6.9
Chemicals	229	1764	1766	457	4234	6.9
Textiles	314	163	749	193	1419	2.3
Paper	371	267	718	109	1465	2.4
Bricks	217	121	315	22	675	1.1
China	17	196	411	77	701	1.1
Cement	817	182	199	76	1274	2.1
Other trades	343	205	2174	425	3147	5.2
Transport	27	—	12403	446	12876	21.1
Railways	*23*	—	*89*	*446*	*558*	
Road	—	—	*9974*	—	*9974*	
Water	*4*	—	*432*	—	*436*	
Air	—	—	*1908*	—	*1908*	
Domestic	5328	4815	1668	3114	14925	24.5
Public services	708	512	1886	449	3555	5.8
Miscellaneous	225	571	1209	978	2983	4.9
Total	12807	11007	29340	7849	61003	
% by fuels	20.9	18.0	48.1	12.9		

This table shows the consumption of different fuels by various consuming sectors. The figures (for 1973) do not include the energy 'over-
heads' of the fuel industries. A further 26.9 thousand million therms are
lost in conversion and distribution of energy: in all, the UK consumed
92.7 thousand million therms of primary fuels in 1973.

SOURCE: as for table 1.4.

cement. Glass, on the other hand, took 7 to 9 per cent of its delivered price to pay for its energy consumption.

The research basis

It is impossible to overestimate the importance of correct allocation of R&D resources. And we can allocate wisely only if we
look very closely at the way in which energy is used, as well as
where it comes from. Thus these statistics reveal the facts of
life that will determine the future of the energy system in the

developed countries. Clearly we can delve deep into the numbers in search of ever more refined detail of the workings of the energy system; but energy analysis can become an addiction. Failing funds to develop hardware and new ideas, there is nothing easier than to equip the researcher with a set of statistical tables and set him off in search of new revelations of the complexity of the energy system. The growth of 'paper research' on energy resembles the earlier booms in various social sciences. Ultimately you end up with a lot of theories, but little concrete information.

There is no doubt that the energy system needs to be thoroughly studied before there is a mad scramble to spend money on R&D. One area of analysis that is much in vogue is energy accounting. Here the aim is to describe a process or an industry in terms of its energy flow. And this flow is more than the direct purchases of fuel by the industry. The energy accountant is also searching for those hidden energy inputs that went into the industry's material inputs, for example. Not only does this technique make it possible to make a more detailed assessment of the impacts of a changed energy situation on an industry, it also enables the energy industry to assess its net contribution to energy supplies. For example, nuclear power tends to be viewed in terms of its uranium input and its electricity output: there are other inputs that have to be taken into account. These include the energy costs of building a power station and the energy costs of the whole fuel cycle, from mining through to waste-handling. This assessment is particularly valuable when two programme options are being considered. For example, two nuclear power projects may appear to have similar economic balances, but the energy balance may favour one option over the other.

One final input on energy consumption should be fed into any R&D policy, and this is a projection of future consumption. We cannot know how much energy will be used in ten years time, but we can make some guesses. There can, however, be no such thing as a single projection of anticipated future energy consumption. There are so many variables that any single projection will soon turn out to be wrong. Any forecast of a country's energy consumption should be accompanied by a list of the assumptions that went into it. The list would include forecasts of national economic

Table 1.6. *Projections of energy consumption by
OECD countries (million tonnes coal equivalent)*

	1972	1980			1985		
		Base	$6	$9	Base	$6	$9
Coal	956.9	1115.4	1283.6	1271.3	1358.6	1497.4	1480.1
Oil	2739.0	4157.6	3492.6	3105.4	5101.6	4015.6	3502.3
Natural gas	1063.6	1360.9	1419.3	1551.9	1582.6	1710.0	1778.3
Nuclear	49.7	437.0	465.1	465.1	938.9	1080.1	1080.1
Geothermal and hydroelectric	138.4	168.4	176.4	177.9	190.6	220.1	230.7
Total	4947.6	7239.3	6837.0	6571.6	9172.3	8523.2	8071.5

The 'base' case is built on projections made in 1973 before the oil
'crisis'. The other projections are for oil at $9 (in 1972 prices) and $6
a barrel. The OECD report gave the projections in million tonnes of oil
equivalent; these have been converted to coal equivalent using the
conversion factor from the report (1 mtce = 0.7 mtoe).
    SOURCE: *Energy prospects to 1985*, OECD.

growth and world oil prices, among other things. At the beginning
of 1975 the OECD issued a report on energy prospects to 1985.
In this the OECD presented various estimates of energy consumption over the next ten years. These projections are based on
various forecasts of oil prices (see table 1.6).

Chapter 2

Reserves and resources

Energy reserves are not like money in the bank. There is no way of knowing exactly how large the world's energy reserves are. We will not know that a particular fuel has run out until production has ceased and the search for new reserves has been abandoned. Before that happens alternative sources of energy will have to be developed if society is to survive. If ever 'independence day' comes, and it could come in the not too distant future if we succeed in tapping renewable sources of energy such as solar power and the almost infinite reserves offered by thermonuclear fusion, we can forget about the world's energy reserves. Until then we must work with the estimates of energy reserves that we already have.

There are no universally accepted estimates of the energy reserves in every country, and any estimates of world reserves have to be treated with extreme caution. But nations have to base their energy policies on something, so they each have their own estimates of the energy resources within their jurisdiction. At the same time there are the more academic and theoretical estimates of energy resources made by less cautious souls who do not have to formulate national policies, and who want to get some idea of how much energy may be locked up in the Earth's crust. Beware of attaching too much weight to anyone's figures. Governments can distort estimates of reserves for political reasons. And while national geological surveys prepare many of the estimates of coal reserves, it is often left to the oil companies to assess oil and gas reserves. Over the years the oil industry has shown that it is quite capable of distorting the numbers it reveals. Even academics have been known to distort their calculations (if only by selective interpretation) to support their theories.

No government will ever say that its estimates of 'proven' reserves are definite numbers that will not change as time goes by.

What they will say is that it makes no sense to plan a country's economy on the basis of anticipated reserves if there is a high probability that those reserves just might not be found. The reliability of reserves estimates depends upon how they were arrived at. Purely geological estimates are less reliable than estimates derived from instrumented surveys including, for example, seismic measurements. The most reliable estimates are made by making contact with the resource in question; and this is what the oil companies do when they make test drillings. Even these are not completely accurate, you have to complete a whole series of drilling tests before you can say with any certainty just how much oil, or coal, or uranium ore, lies in a particular formation. This drilling can be expensive business – you certainly do not do it just to prove that something isn't there, in other words you do not drill on hopeless prospects. Just to make things difficult, there are, of course, cases where someone has drilled into a 'hopeless' piece of ground only to discover a large oil field.

At any one time we can rely only on 'proven' reserves, where detailed exploration has given us some certainty that a measured amount of material is in-place and awaiting exploitation. Beyond this we can talk with varying degrees of conviction about resources we *expect* to find as a result of further exploration. Governments are unhappy when forced to adopt policies that are based upon too many unknowns. They will not, therefore, embark upon drastic changes in their energy systems if these depend upon anything as unreliable as yet-to-be-discovered energy reserves. This partly explains why the British government is slower than the oil industry to upgrade its estimates of North Sea oil and gas reserves. The bankers, oil companies, and others whose business it is to speculate on new discoveries have to be a little bit ahead of everyone else if they are to make the most out of the North Sea.

The level of energy reserves does not depend solely on knowledge of the presence or absence of particular materials in a given location. Uranium, for example, exists in many minerals, and even in seawater (3 parts in 10^9, by weight); but try to extract that uranium and you soon learn that there are factors other than the presence of the element that determine the world's uranium 'reserves'. Perhaps the most obvious secondary factor influencing the level of reserves is technology. A rich coal vein, for example, is useless if its location makes it unexploitable with modern technology. In this case the coal might be under the sea bed – the companies drilling in the North Sea have found coal in their search for oil and gas. While we are rapidly becoming adept at extracting oil from under the sea bed, we have yet to come up with technology that allows us to dig coal if it is hundreds of miles offshore in deep water.

As well as geology, our true energy reserves depend on manpower, money, and 'environmental' factors, such as the availability of water. If the massive oil shale reserves of the US, for example, can be exploited only if massive quantities of water are available, as is explained in chapter 4, the shale could be totally useless if there just is not any water. One final, and perhaps insurmountable constraint is the amount of energy that has to be consumed in exploiting new energy reserves. If you have to put more energy into extracting fuel than is available in that fuel, then there is not much point in exploiting the reserves. This state of affairs could be the case for uranium in seawater unless less energy intensive extraction techniques are found.

We will look here at the more conventionally understood energy reserves, but if we are to arrive at any meaningful conclusions about the way in which energy technologies might develop, we must also delineate some of the other factors that determine our ability to supply the energy we need.

Fossil fuels

The newest compilation of official estimates of world energy reserves was prepared for the 1974 World Energy Conference (WEC). This important anthology was built up from figures submitted by national organisations. While this does not make the figures any more accurate than less official estimates it does mean that they are more likely to be used in formulating national energy policies. Dr V. E. McKelvey, Director of the US Geological Survey, underlined the importance of such reserves estimates in December 1974 when he pointed out that the US's estimates of the oil and gas potential of the Atlantic Outer Continental Shelf are still very tentative: 'The data we have acquired is far short of what we could wish for and expect eventually to have. It must,

Table 2.1. *World energy reserves, as estimated in 1958*

	Coal equivalent (thousand megatonnes)	% of total
Coal and lignite	4400	91.8
Petroleum	150	3.1
Natural gas		
With petroleum	90	1.9
In coal measures	45	0.9
Oil in oil shale	45	0.9
Peat	65	1.4
Total	4795	

This study of world energy reserves added that 'probably not more than one-third to one-half of the reserves of coal and lignite could be economically mined'. Nuclear power was left out because 'no one can at this date say how accurate are the forecasts' of the growth of nuclear power.

SOURCE: *The efficient use of fuel*, 2nd edition, HMSO, 1958.

nevertheless, provide the basis for at least a tentative estimate of the oil and gas potential of the Atlantic Shelf.' As he said, 'there is a much better case for expecting large quantities of oil and gas to be found under the Atlantic OCS than there is for condemning this region as unproductive, but we shall never really know until it is tested by the drill'. But without these early guesses the US cannot arrive at a policy on this drilling. For example, it does not know how much to charge the oil companies for the privilege of drilling into this new region. Tentative estimates should therefore help shape national energy policies and R&D programmes, and play a significant part in shaping future energy technology.

Before we take a look at the WEC statistics, a quick glance at some older estimates (see table 2.1) shows just how they can vary over the years. Twenty years may seem like a long time in terms of modern technology, but it is a fleeting instant in the life of energy reserves and a short interval in terms of the time it can take to bring a new idea out of the laboratory and into industrial practice.

The 20-year old estimate is surprising in that its figure for petroleum reserves is quite similar to the WEC estimate of

Table 2.2. *World coal reserves: the world's recoverable reserves of coal could last many years at current consumption rates*

		Reserves			
	Production (kilotonnes)	Recoverable reserves (mega-tonnes)	'Life' (years)	Total (mega-tonnes)	Total resources (mega-tonnes)
USSR	694000	136600	197	273200	5713600
China	410000	80000	195	300000	1000000
Asia	176052	17549	100	40479	108053
US	509930	181781	356	363562	2924503
Canada	17600	5537	315	9034	108777
Latin America	10980	2803	255	9201	32928
Europe	1117680	126775	113	319807	607521
Africa	60442	15628	259	30291	58844
Oceania	85290	24518	287	74699	199654
Total	3081974	591191	192	1420273	10753880

Total resources are an order of magnitude greater than the recoverable reserves.

SOURCE: *Survey of energy resources 1974*, World Energy Conference.

'published proved' oil reserves. Since the older estimate was made crude oil production has been around 50 thousand megatonnes.

Much as they make interesting reading, historical estimates of energy reserves tell us little more than that it is folly to trust them too much. This is a clear warning that we should not place too much faith in today's estimates. But they are all that we have to go on for the time being. Because the official pronouncements of the world's energy producing nations are enshrined in the WEC estimates, this compilation must carry more weight than others.

The first shock that comes from the numbers in table 2.2 is the disparity between total resources and recoverable reserves. Estimates of coal resources are taken more seriously than other resource estimates, because coal geology is relatively simple. 'Resources' refers to the amount of material that might be exploited 'within the foreseeable future', as the WEC survey put it. 'Reserves' refers to the amount of material that has been accurately assessed and determined to be exploitable given current

Table 2.3. *World oil reserves: these are small
and unevenly distributed*

	Proved recoverable reserves (megatonnes)	Production (megatonnes/ year)	Reserves/ production ratio (years)
Africa	12 848	272.4	47
Asia	53 972	1021	53
Europe	1 394	35.43	39
USSR	8 138	394	21
Canada	1 075	69.41	15
US	5 569	436.15	13
South America	7 599	218.3	35
Oceania	229	15.4	15
Total	91 525	2493	37

The reserves/production ratio, found by dividing proved reserves by current production, shows that oil reserves can be measured in decades rather than centuries.
SOURCE: as for table 2.2.

economic and technological conditions. What really matters to the consumer is the amount of 'recoverable reserves', which is the amount of material that can be recovered under current economic and technological conditions. It is impossible to extract all the coal from a coal field, or oil from an oil field. Something like half the coal in an underground coal field, for example, may have to be left where it is to stop the roof from falling in. And it is not yet possible to extract more than a third of the oil in many oil fields.

Comparison of the current level of consumption with the recoverable reserves shows just how significant the reserves are. In the last chapter we saw that in 1970 world consumption of solid fuels was 2415 megatonnes. Thus, if coal consumption continued at today's levels the recoverable reserves would see us through another couple of centuries before coal runs out. And if the 'total resources' can be tapped to any extent, coal could last for a very long time. But, on the other hand, a massive switch away from oil as supplies diminish, and failure to develop nuclear power, would soon see coal consumption making appreciable inroads on the reserves. We should remember this when we talk of making oil and gas from coal, and of building more coal-fired power stations.

Crude oil reserves are, as we can see from the next WEC table (table 2.3), in nothing like so rosy a situation as coal. Proven reserves are less than a third those for coal, and there is little prospect of the 'total resources' of oil reaching anything like those of coal. And, what's worse, oil consumption is significantly higher than coal consumption. On top of this there are geographical differences between coal and oil. A third of the recoverable reserves of coal are in the US; just 8 per cent of the world's oil reserves are in North America. (The US has around 6 per cent of the world's oil reserves.) Thus the world's major energy consumer has large coal reserves but much smaller oil reserves. Clearly this state of affairs will have an impact on the energy technologies that will be adopted over the next few years.

Natural gas is something of a confusing fuel (table 2.4). At one time it was next to worthless and was an undesirable waste product that came with oil. Before new transport technologies involving cryogenic storage were developed much natural gas was 'flared' (burned off) at the well head. Where the gas was found near potential consumers, in the US for example, it soon became a valuable fuel. As far as Britain goes, the North Sea first revealed itself as a source of natural gas. As far as gas reserves go, the situation is similar to that for oil. Once again the US, which has a sizeable natural gas market, has limited reserves.

Nuclear materials

Uranium reserves belie the previous suggestion that governments would not commit themselves to a course of action that depends upon the discovery of anticipated reserves. The world has embarked upon rapid development of nuclear power, and yet we cannot be sure that the uranium reserves are there to sustain this growth. This apparent departure from sane planning owes much to the nature of the nuclear business. Unlike coal and oil, uranium was not a valuable commodity until very recently. (At the same time, its geology is totally different from that of the fossil fuels.) There was, therefore, no incentive to look for it. Uranium is a relatively abundant element. Unlike fossil fuels it is not formed

Table 2.4. *World gas reserves: while some regions have fairly healthy gas reserves, in terms of current consumption, the largest consumer, the US, is in a desperate situation*

	Proved recoverable reserves (cubic km)	Production (cubic km/year)	Reserves/ production ratio (years)
Africa	5709	38.82	147
Asia	12241	151.2	81
Europe	4513	179.5	25
USSR	17136	212	81
Canada	2576	73.11	35
US	7556.5	637.5	12
South America	1591	70.57	23
Oceania	693.5	3.73	186
Total	52532	1389	38

SOURCE: as for table 2.2.

as a result of any sequence of biological and geological accidents. It is present in low concentrations in many minerals and in seawater. Given that the search for uranium has not been comprehensive or particularly vigorous it seems reasonable to assume that we will find further reserves of uranium ore.

Because they have been important over a relatively short period, and perhaps also because the nuclear industry is a high technology energy industry, uranium reserves are enumerated in a more sophisticated way than those of other energy resources. Whereas coal and oil are quoted in terms of 'recoverable' reserves, with the definition of recoverable a tenuous thing, uranium is listed according to the anticipated price of mining it. The OECD's Nuclear Energy Agency (NEA) has become the accepted authority on uranium reserves. In 1973 the NEA published a joint report from its own experts and the International Atomic Energy Agency. This was the latest in a continuing series of reports (see p. 219).

Uranium prospecting has been an up and down business. At one time it was going to spark off a new American 'gold rush'. Between 1949 and 1954 the US government sold 100000 copies of a booklet entitled *Prospecting for uranium*. The slowness of

Table 2.5. *Nations with major uranium resources*

	Reasonably assured resources up to $26/kg uranium (tonnes)	Total uranium resources (tonnes)
US	329267	2041156
Canada	185799	716984
Sweden	—	308381
South Africa	202000	298004
Australia	120949	160049
France	34850	85000
Niger	40000	80800
India	—	61862
Colombia	—	51000
Argentina	12665	38590
Gabon	20400	30240
Rest of Europe	21834	73863
Rest of world	16710	78023
Total	984474	4023948

SOURCE: as for table 2.2.

growth of the nuclear industry, together with a large government stockpile of uranium, has held back the search for new reserves. Indeed, the NEA report said that: 'During the three years since the last report, uranium exploration in the United States has declined.' During those three years – 1970 to 1972 – uranium reserves that can be mined at less than $10 per lb of U_3O_8 ($26 per kilogramme uranium) grew by 98 kilotonnes. In the same period 31 kilotonnes of uranium were mined, resulting in a net addition to the US reserves, taking them from 192 kilotonnes uranium on 1 January 1970 to 260 kilotonnes uranium on 1 January 1973. In the same interval estimated additional reserves rose from 460 kilotonnes to 540 kilotonnes. While this discovery rate is not high enough to sustain growth of nuclear power for long, it is adequate for more immediate needs.

As we have said, there is no reason to suspect that more uranium will not be found, but the search will have to be stepped up if there is not to be a shortage of uranium in the not too distant future. According to the NEA report: 'Annual demand for uranium is expected to establish itself in the region of 60,000 tonnes uranium by 1980 and almost double this figure by 1985. No shortages of uranium supply are to be expected in the 1970s. However, the rapid growth in demand in the coming decade cannot be satisfied on the basis of existing uranium exploration levels. Given the necessity of a lead time of about eight years between discovery and actual production, it is therefore essential that steps be taken to increase the rate of exploration for uranium so that an adequate forward reserve may be maintained.'

The low price of uranium, which is now going up, has been a constraint on prospecting. But with the price of U_3O_8 rising from \$7.65/lb for 1974 deliveries to \$36/lb for 1980 deliveries the incentive is stronger for new exploration. Indeed, the US Atomic Energy Commission's 1974 report on uranium exploration said that the industry planned to spend \$150 million on uranium exploration in 1974/5, although this might prove difficult given a possible shortage of drilling equipment. Other countries were also active in uranium exploration and reserves assessment. Sweden, for example, spent Sw.Kr 2.6 million during the 1973/4 financial year on the production of uranium from shale. Sweden has large reserves of uranium bound up in low grade shales.

The appeal of uranium exploration may depend upon the price of uranium, but the same is not really true of the appeal of nuclear power. The cost of uranium is but a small part of the cost of running a reactor. Indeed, uranium prices could be ten times higher than the \$26/kg price tag and nuclear power would still be competitive with electricity generated in fossil-fuelled power stations. Thus nuclear power could survive rapid rises in the price of uranium. If these occur there will be a mad rush to look for new deposits, and exploit some of the low grade sources that are now ignored. One low grade source that is tapped, but that would be exploited even more at higher prices, is the waste material that is thrown out after other materials have been taken from an ore. South Africa, for example, already extracts uranium from the tailings, the leftovers, from gold mining. These tailings have approximately 0.2 per cent uranium in them.

The amount of energy 'available' in a tonne of uranium depends very much on the technology of the reactors employed. Thus it is inadequate to talk about uranium reserves merely in tonnes of metal. In a 'thermal' reactor of the type now being built in increasing numbers throughout the world, the energy that is tied up in uranium-235 can be extracted. But natural uranium consists of 0.7 per cent uranium-235; uranium-238 makes up the rest of the metal and is relatively useless in a thermal reactor. In a fast neutron reactor (see chapter 6) the uranium-238 isotope can be more efficiently converted into a fissile fuel, thus increasing the energy content of the original uranium by a factor of 50 or more. Thus the 'life' of the uranium reserves is greatly influenced by the technology employed. Uranium reserves will last a lot longer if, and it is a big 'if', the fast breeder reactor can be successfully introduced.

Another factor influences the supply of useful uranium. While some reactor designs – most notably the Canadian Candu system – can consume uranium in its natural form, most of the reactor types now in operation or under construction are fuelled with uranium that has had its uranium-235 content boosted from its natural 0.7 per cent to something like 2 or 3 per cent. This enrichment process is both capital intensive and energy intensive. The established enrichment plants that provide nearly all the world's enrichment capacity are huge, and they gobble up electricity at an amazing rate. New enrichment technology – the gas centrifuge – can, however, cut the energy required for a given level of enrichment by 90 per cent; and gas centrifuges can be built in smaller units that can be scaled up as demand for enrichment capacity grows (see chapter 6). There could be a shortage of uranium enrichment capacity even if there is no shortage of uranium. This is, therefore, another factor that we must take into account when we are looking at the way in which the energy system could develop.

This then is the situation with the reserves of 'conventional' fuels. There is no immediate shortage of energy. But there are serious doubts about our ability to meet a growing demand for energy if that demand continues to grow at anything like the rate of recent history. There are signs that higher prices have slowed down the inexorable growth of energy consumption; but the 'forward reserves', especially of oil and gas, will not take us very far. As we shall see in the next chapter, energy research and development can be a slow business. It could take a long time to switch from the 'classical' fuels to the more radical energy systems whose reserves and resources are outlined below.

Unlimited fusion

Sitting on the horizon, a horizon that sometimes seems to recede as fast as we approach it, is an energy system that just might be the final solution to all of the world's energy problems. Thermonuclear fusion drives the stars and the Sun, it also powers the hydrogen bomb. As yet we do not know if it can energise our society. Just how much remains to be done before fusion power becomes a reality we will see later (in chapter 7), but a look at the relevant reserves shows just why it makes sense to spend a lot of effort and money on a technology that sometimes seems glued to the drawing board.

Fusion takes place when two light nuclei collide with enough force to make them unite into a single new nucleus. Energy is generated in the process and released as energetic particles. If the army of scientists and technologists working on the problems of fusion are successful, this energy could be tapped and used to generate electricity. The fuels for the first fusion reactor will probably be deuterium and tritium – two isotopes of hydrogen. Other thermonuclear reactions are possible – the deuterium–deuterium reaction is very appealing – but they seem unlikely candidates for the first round of reactors because they pose even greater problems than the deuterium–tritium reaction.

Deuterium supplies are no problem. There is some deuterium in all water (about 1 part in 5000). It does not make sense to talk about deuterium reserves in units as low as millions of tonnes of coal equivalent. A more convenient unit is Q, which is 10^{18} British thermal units (Btu). A million tonnes of oil have a little more than 10^{12} Btu available energy. And the world's energy consumption is now something like 0.2 Q a year. If, as might one day be possible, fusion power stations can be fuelled by deuterium alone, then the world's energy reserves would be 10^{10} Q. When that lot runs out mankind will be worrying about things other than a shortage of energy.

In the meantime, tritium 'breeding' rather than deuterium

reserves will be the constraint on fusion power. Tritium is not something that nature can provide. There are no tritium 'reserves'; it has to be made, and the most attractive way of doing this is by surrounding a fusion reactor with a blanket of lithium. Deuterium–tritium reactions produce neutrons; when these are absorbed in lithium a nuclear reaction can take place as a result of which tritium is generated. Thus lithium reserves are the constraint on deuterium–tritium reactors. Geological estimates of land-based reserves suggest that lithium ores exist in reserves equivalent to around 1000 Q. Thus deuterium–tritium reactors could keep society going at today's energy consumption rate for several thousand years, which should give us time to come up with a replacement energy system.

It would be comforting if this were the true extent of the reserves critical to a fusion economy. Unfortunately, there are other materials we have to take into account. Fusion reactors, as currently envisaged, would require some sophisticated materials which have nothing to do with the fuel cycle. For example, it seems likely that superconducting magnets will have to be used in a fusion reactor. The superconducting metals now available consist of alloys of certain exotic metals. These will almost certainly be in short supply long before fusion fuels show any sign of running out. It is important to know what these limiting factors on fusion development might be when an R&D programme is being formulated. There will inevitably be several paths that fusion research can take; it might be profitable to take a path that, although it poses greater technical problems, has less severe materials challenges.

Fusion power is the ultimate 'consuming' energy system. It offers us access to the largest resources stockpile. It is not, however, the only way to meet our future energy needs. We can also tap the 'renewable' resources, that is those Earthly inputs that nature regularly restocks for us. Solar radiation is the most obvious of these. At the same time we could tap tidal power and wave power. We already exploit hydropower to a considerable extent – so much so that in most of the developed nations technically feasible and socially acceptable sites for hydroelectric power have been all but used up. Yet another energy source that is of considerable magnitude is the heat of the Earth's core – geothermal energy. And we should not forget the wind. Windmills used to be very popular, until cheap fossil fuels came along to do the job more efficiently and predictably.

Nature's resources

You can easily prove that all this talk about an energy crisis is meaningless by adding up the amount of energy available to us from natural sources. Unfortunately, if you add up the energy of the winds, tides, waves, rains, and Sun, all you end up with is impressive documentation of the power of nature. M. King Hubbert's influential article on the energy resources of the Earth (*Scientific American*, September 1971) tells us that: 'The total solar radiation intercepted by the Earth's diametric plane of 1.275×10^{14} square metres is 1.73×10^{17} watts.' And on geothermal energy King Hubbert says: 'the average rate of flow of heat from the interior of the Earth has been found to be about 0.063 watt per square metre. For the Earth's surface area of 510×10^{12} square metres the total heat flow amounts to some 32×10^{12} watts.' On top of this: 'The energy from tidal sources has been estimated at 3×10^{12} watts.'

Clearly the Sun is the most prodigious supplier of energy. King Hubbert's estimate of the solar radiation that reaches Earth converts to something like 5000 Q per day, or 25 000 times as much energy as we consume each year. As the proponents of solar energy so tirelessly point out, and King Hubbert confirms, 'the area required for an electric-power capacity of 350,000 megawatts – the approximate capacity of the US in 1970 – would be 24500 square kilometres, which is somewhat less than a tenth of the area of Arizona'.

It isn't every country that has such large tracts of well-situated land where it can contemplate building solar 'farms'. The situation in Britain was spelt out in a report prepared by the UK section of the International Solar Energy Society (ISES). According to this report: 'At Kew the mean power in winter on a horizontal plane is 0.048 kW/sq.m, while in summer it rises to 0.163 kW/sq.m. The mean annual power in America is 0.180 kW/sq.m and in Australia 0.200 kW/sq.m, so the difference between temperate and more tropical latitudes is not as great as is often imagined.

The greatest differences are in winter.' The report goes on to say that the whole of Britain's current electricity need could be met if 1 per cent of our land area could convert solar energy to electricity at 10 per cent efficiency. It may be that even in Britain, where the inhabitants sometimes seem to revel in their lack of sunshine, solar energy could be put to good use in some locations and for some purposes. When the oil crisis hit Britain no one really knew how much energy could be taken from renewable sources. As a result of the new concern for energy supplies, the UK section of ISES set up a study group to reassess the country's solar energy resources.

The World Energy Conference has also looked at renewable reserves. It too steers clear of over enumeration of what are essentially meaningless quantities, contenting itself with broad order-of-magnitude estimates. For example, the WEC tells us that ocean thermal gradients – the temperature differences between an ocean's surface and its lower depths – offer a 'very large' amount of energy. Indeed, 'the total energy available from the Gulf Stream along the southern Atlantic coast of the United States is much greater than present US energy consumption'. The main conclusions of the WEC survey can be summarised as follows:

Geothermal	Relatively limited, but very important in some areas
Tidal	All possible sites, if utilised, could supply up to 13–20 gigawatts, with an average annual output of around 175 000 gigawatt hours of electricity
Wind	Extensive use 'could supply many times the energy available from tidal or natural steam geothermal resources'
Solar	Could be a 'near infinite' supply of energy

In all, the renewable energy resources have, so the WEC survey tells us, 'a capacity far in excess of present or future world needs for energy'. And yet they do not play a significant part in today's energy system even though natural sources provided the energy that helped man clamber out of the caves and into the sky-scrapers. During the past few centuries there has been little incentive to employ the renewable resources even though the fuel is 'free'. Non-consumable energy systems have invariably demanded a significantly higher capital investment than fossil and, more recently, fissile fuels. At the same time, more often than not there simply was not the technology available to harness the 'elements'. Fuel prices are now rising to levels that make it economic common sense to invest money to reduce running costs. This, and the fact that few people now doubt that fossil fuel will eventually run out or become ruinously expensive – as time goes by we will eat into the 'stocks' of these fuels making it harder and harder to dig up enough fuel to meet the world's needs – have engendered a new and more urgent interest in the renewable energy resources.

Hidden limits

This thumbnail description of the world's energy reserves situation underlines our dependence on factors other than the physical limit of buried resources. It is not only in the reserves and resources figures that we must seek clues as to the future shape of the world's energy system. The estimates of reserves, when related to consumption, tell us why there is so much interest in new energy technologies. Oil, the most popular energy form, is disappearing fast. It is backed up by the smallest reserves (in terms of life expectancy) and the reserves are in the wrong countries (consumers and suppliers are different people). But we cannot even rely on this gloomy prognosis when we determine R & D priorities. There are some knowledgeable and optimistic experts who expect the shift from onshore oil exploration to offshore exploration to reveal at least as much oil as has already been found. But this will depend upon significant technological advances springing from a sizeable R & D effort. The technical difficulty and high cost of bringing ashore any oil that is found way under the sea bed almost certainly mean that any oil that is found will be far more expensive than even today's high cost oil. Thus the days of the 'monoculture', with an energy system dominated by oil, are over and other energy sources will compete to do the jobs they are best suited to.

A pessimist could have a field-day with the world's energy situation. Oil's future is doubtful. There is enough coal around to keep us going for some time, but the technology needed to extract and utilise this fuel is still being developed. Uranium reserves could come under pressure within a very short time of nuclear power reaching commercial maturity. Fusion power could bale us out of any energy predicament for a very long time; but it will be some time before we can be sure of this, and even longer before we can begin to build fusion power stations on any scale.

If this scenario is not bad enough, we can heap on the misery by taking a closer look at some of the other reserves and resources' factors which determine our future energy supplies. Perhaps two of the most comprehensible limitations are potential shortages of water and money.

Money is needed if we are to build new coal mining capacity, for example. One coal expert has said that: 'The size of the task of creating the required level of new and replacement capacity is such as to raise doubts whether, in fact, the various resources needed to support such a programme are indeed available, and if so, whether they can be deployed at the right places and in time.' It costs somewhere between £5 and £20 to add a tonne a year to the world's coal mining capability. We might need to establish an annual production of 10000 million tonnes of coal a year by the year 2000, something like four times today's output. Eighty per cent of this new capacity might be in underground mines, the rest in open-cast mines. Over the next 20 years the underground capacity might cost £53000 million to establish (at 1974 prices and assuming an average cost of £12 per tonne of new annual capacity). Open-cast mining costs less to implement – say an average £4 per annual tonne – but it could cost £4400 million to establish enough capacity to meet the anticipated demand for coal. Thus the total capital requirements for new coal mining capacity could be something like £57400 million, which represents an annual investment rate not far short of £3000 million. All this money will provide us with piles of coal. Yet more money will have to be invested if this coal is to be put to good use. If it is to be burned in power stations, each £1 invested in setting up new coal mining capacity may have to be matched by £10 for generating equipment.

Nuclear power and offshore oil will be equally costly to develop. Electricity utilities in some countries have aready experienced great difficulties in obtaining the capital they need to build new power stations. In the US, for example, the utilities wanted to order a rush of new nuclear power stations in the wake of the oil crisis, but the money just was not available. Indeed, work had to stop on at least one power station simply because the utility involved had run out of cash. Only the fall in demand that came about when energy prices rose so dramatically has saved many utilities from almost certain shortages of generating capacity in a few years time. Other energy systems could be even more costly to develop than nuclear power, coal, or offshore oil. So even if there is plenty of energy available in the Sun's radiation, for example, we cannot be sure that our money resources will be up to implementing any new technology.

Chapter 3

Back to the drawing board

An energy doomsday may not be looming over us, but we do not have immediate access to the technologies that can translate notional reserves and resources into real energy supplies. So complicated are many of the technological problems that we cannot guarantee that the requisite technologies can be developed in time to stave off a future shortage. Only an extended energy R&D programme will bring us the hardware and systems that are essential to a continued flow of energy. And this R&D effort must not be abandoned as soon as the immediate shortages are over.

Part of the difficulty lies in the distorted energy situation of the past two decades. Thanks to the low cost of obtaining massive and rising supplies of oil, this energy carrier has all but crippled the world's coal industries, for example. At the same time inexpensive oil has made new energy technologies an unattractive investment. To a certain extent we can even blame the continuing low price of oil (a price that has fallen significantly in real terms over the past twenty years) for the poor performance of nuclear power. Ten years ago scientists were predicting that within a very short time uranium would take over as the most popular fuel for electricity generation. This did not come about as quickly as was predicted, partly because nuclear technology turned out to be more difficult to develop than was initially anticipated, but also because electricity utilities leapt at the ever falling price of oil and ordered as much oil-fired generating capacity as they could. Now, of course, these utilities are very worried; but it will be a long time before they can reduce their heavy dependence on oil. (The life of a power station is more than thirty years.) And yet oil will almost certainly be the fuel that will show the first signs of exhaustion, perhaps even before some of today's new oil-fired power stations are past their useful lives. Thus the world has suddenly discovered

[33]

that it depends for most of its energy on the least secure and most short-lived energy carrier. And no other energy system is well enough developed to fill the gap. It is, therefore, no surprise that energy R&D is increasingly fashionable. We are trying to make up for twenty years during which nuclear power was the only technology to see significant R&D effort. The OECD study of energy R&D published at the beginning of 1975 pointed out that: 'Of all the energy sources, nuclear energy constitutes the main R&D objective in most countries. The proportion devoted to it, which in the case of a few countries is slightly lower than 50 per cent, ranges in all the others from 60 per cent to 85 per cent of the total energy R&D budget.'

The situation would not be too bad were it not for the size of the energy business. The massive investment in a system based heavily upon oil makes it very difficult to alter the direction in which the energy business is heading. You cannot scrap overnight the huge investment in oil-fired power stations and an oil-based transport system. This is not the only inertia that has to be overcome if the world is to switch away from oil and on to alternative energy carriers. Because energy R&D has been something of a wasteland for so long, it will be some time before new energy technologies can be developed to such a stage that they are commercially viable. There will then be a period of growth during which the new technology will begin to make an impact on the world's energy system.

Technology transfer

Science does not make its way out of the laboratories and into the showrooms by some magically simple process. The processes of discovery and innovation are complex enough; add to that the difficulties of industrial capability and public acceptability, and you begin to see why it takes so long to introduce a new energy system. Take the case of nuclear power: we can trace the growth of the nuclear energy industry back to basic scientific research carried out in the late 1930s and early 1940s. During that period the first experimental nuclear reactors were built and operated. They were nothing like the sophisticated and enormous constructions we see today, but their potential was clear and governments responded to this by pumping money into nuclear research and development. Admittedy, there was also a military component in many nuclear programmes – Britain's leadership in the early days of commercial nuclear power owed much to military ambitions – but there was also a growing conviction that nuclear power would be the energy system for the future. The promise was not that easily realised; even today, more than 30 years after the early experiments, the atom makes a relatively minor impact on the world's energy system.

Low oil prices helped impede the introduction of nuclear power. So much so that it was only after the 1973 Middle East war that reactors became the number-one choice of fuel for a new power station. Before then there had been something of a battle between oil and uranium – with coal playing a minor role where it was appropriate – whenever a utility wanted to build a new power station. A return to this situation seems very unlikely. Even if the price of oil does fall, nuclear power will almost certainly remain competitive; and because fuel prices play a relatively minor part in determining the costs of electricity from a nuclear power station, uranium will be viewed as a relatively stable energy source in comparison with oil with its proven uncertainty. (Uranium has the added advantage that no single political bloc can control supplies in the same way that the oil producing nations can.) Thus many countries are now predicting a more rapid rise in the nuclear share of electricity generating capacity than they were just a few years ago. This should take place despite the temporary difficulties that have left electricity utilities so short of money during the period of rapid inflation following the sharp rise in oil prices. The near bankruptcy of many utilities has caused them to postpone or even cancel new nuclear plant.

A quick look at the development of nuclear power shows that, taking 1945 as the start of the nuclear era, it was nearly 30 years before nuclear power really caught on and perhaps it will be 50 years before uranium makes a sizeable impact on the energy market. Thus the timetable for developing and introducing new energy technologies can be long and drawn out; and it will take longer to bring in new technology as the system into which that technology is to be introduced grows.

If it takes a further 50 years to introduce fusion or solar power, for example, and to see them taking anything but a very small share of the market, then today is by no means too soon to start taking R & D very seriously. In fact there has, as we shall see later, been a significant programme of research on fusion for 20 years or more. And while other energy technologies would probably take less time to come to something, it is difficult to see how any new energy system could begin to become widespread enough to affect the consumption of other fuels before the end of the century. Even without the added complication of an R & D programme the lead times for new energy projects are sizeable. The report *US energy prospects: an engineering viewpoint* from the US National Academy of Engineering lists some typical overall project times, giving the delay between giving the go-ahead to a project and the production of energy resources:

Coal-fired power plant	5–8 years
Surface coal mine	2–4
Underground coal mine	3–5
Uranium exploration and mine	7–10
Nuclear power plant	9–10
Hydroelectric dam	5–8
Production of oil and gas from new fields	3–10
Production of oil and gas from old fields	1–3

Clearly a judgement of the priorities for energy R & D should take the likely lead time for the resultant technology into account. The OECD report *Energy R&D* discusses R & D prospects in three categories: short-, medium- and long-term. The report says that 'In the short term, OECD energy needs will be satisfied mainly by the same resources and technologies as exist today.' It adds that 'R & D now being pursued will have a much broader impact in the medium than in the short term'. The short term is taken to be between now and 1985, while the medium term carries us through to the year 2000. As far as long-term R & D prospects (beyond the year 2000) are concerned, 'prospects are obviously much more uncertain'.

The short-term contribution that R & D can make to our energy situation will cover oil and natural gas, coal for electricity generation, and nuclear power. The OECD report suggests that the most important contributions that R & D can make to the oil and gas industry include the following: the improvement of prospecting technologies on and off shore; the improvement of secondary and tertiary recovery methods (see chapter 5); better deep-drilling technologies; and technology that allows the industry to move to deeper waters for oil production. Coal R & D can make its greatest short-term contribution by coming up with new technologies for cleaning the emissions from chimney stacks. Nuclear power needs R & D support to improve the fuel cycle rather than the reactors, which will be based on existing designs.

Coal, oil, and nuclear power have a universal applicability – most countries have an interest in them. There are some energy technologies which, while not so significant in overall terms, could have a real impact in particular areas. The OECD lists five areas where this could be the case: shale-oil in-situ extraction in the US; oil extraction from tar sands in Canada; geothermal energy through dry steam; solar energy for space heating and cooling; and synthetic fuels through pyrolysis of organic wastes.

Medium-term prospects could be far more revolutionary and wide reaching. The gloomy predictions of an oil famine around the turn of the century could come to nothing if an extensive R & D programme allowed the search for oil and gas to move into ocean depths considerably more than 1000 m. Coal conversion technologies could have a similar effect if they can be developed so that oil and gas can be made out of coal. Shale-oil extraction 'could spread to many countries which hitherto have made no systematic effort to discover and assess shale resources'. Nuclear R & D will see the development of the breeder reactor and the high temperature reactor. Non-conventional energy sources could continue their local importance, becoming more versatile and prolific. According to the OECD, 'Among these sources are geothermal energy through hot brines, solar energy for electricity generation, synthetic fuels through bio-conversion of plants or organic waste, wind power, etc.' Electricity should benefit from the arrival of advanced generating cycles, improved transmission, and perhaps some storage potential. In the medium term 'research to improve energy utilisation will be most successful'. And here progress could come in the shape of a slower growth rate for energy demand as new utilisation technologies make energy use more efficient.

Long-term prospects just cannot be predicted at all reliably.

But we can say that if fusion power comes to anything it will have little impact before the year 2000. The OECD also suggests that 'as far as energy conservation is concerned, the biggest long-term impacts of R & D will among others, be found in the transportation and agricultural sectors'.

A balanced portfolio

The days are over when most countries could afford a reasonable, if not large R & D programme in all the promising areas of energy technology. Even the richest countries cannot afford to explore all the paths of energy R & D that might, one day, pay off. Thus it becomes more and more important to ask why research is being conducted, and to ensure that 'frivolous' or less important topics are pursued only as far as is reasonable. Clearly an idea that has an immense potential return should not be ignored, no matter how distant that pay-off may seem. On the other hand, the most easily conducted R & D programme, with few obstacles standing in the way of successful innovation, can be ignored if the returns are minimal. Fusion power and solar energy are long-term prospects. Both require large R & D efforts if their full potential is to be achieved. (Solar energy can be useful now, but scientists and engineers will have to work hard before it takes over more than a fraction of the energy market.) On the other hand, oil's prospects are too dim to justify long-term R & D programmes seeking new ways of putting oil to use if the resultant technologies will require large quantities of oil 50 years hence. That oil may be available, but we cannot be sure of this. Any money would be better spent on exploration and production technologies rather than consumption technologies. R & D that can stretch the world's oil reserves by improving utilisation efficiency would also be valuable, but once again the timetable for the R & D effort is important. If it takes 50 years to develop and innovate a new automobile engine that burns petrol more efficiently, for example, the new technology could be irrelevant by the time it begins to make any impact.

The importance of cost-effective R & D and the realisation that some energy technologies could never come to anything partly accounts for a view of energy R & D less hysterical than we might have experienced had the energy crisis come during the 1960s.

At that time science and technology were still viewed as benign forces: there was also something of a crash programme mentality that led governments to set up huge projects when they wanted something done. Hence the American space programme was instigated when the Soviet Union launched the first space satellite in 1957. In those days money was no object – when NASA sought funds they were forthcoming.

By the time the energy crisis hit home, science and technology had been toppled from their pedestal, thanks to the anti-pollution movement and doubts about nuclear power. Thus the US government did not throw money into the research laboratories quite as quickly as it did when the space programme – which has far less obvious benefits to mankind – was started. Energy R & D budgets in the US did rise after the energy convulsions, but in many cases the groundwork for this extra spending had been laid long before the autumn of 1973. Solar energy research and fusion research, for example, were entering new phases that had nothing to do with the Middle East situation. Throughout the early 1970s solar energy had been under review, with more and more scientists and engineers ready to back it as a significant energy source. Indeed, a joint NASA–National Science Foundation report had called for more R & D on solar energy in 1972. At about the same time fusion research was achieving some significant breakthroughs in the laboratory. Thus the money that flowed into these two areas did not flow as a direct result of the oil embargo. (But we can be pretty sure that the events of the last quarter of 1973 stilled any dissenting voices that might have queried the expenditure of such vast sums had there not been so much awareness of the changed energy situation.) Fusion and solar energy were probably the best placed of the long-term energy prospects – as well as the most promising – but there had been a similar upturn in interest in other 'unconventional' energy sources, such as wind power, geothermal energy, and so on.

If the energy crisis did not immediately unlock the coffers for the scientists and engineers sitting in their laboratories and workshops, it did loosen the purse strings for those groups who wanted to analyse the energy situation. And let us be thankful that for once people were prepared to spend appreciable amounts on some in-depth analysis of what could be done to achieve energy security.

(The same money would have achieved next to nothing had it been spent on hardware or laboratory work.)

The early fruits of the 'paper studies' provided a useful adjunct to the various reports and papers prepared by governments and the working parties that had been set up long before the oil embargo hit home. The Energy Policy Project of the Ford Foundation, for example, cost $4 million and produced a series of some 20 reports on various aspects of the energy situation. Perhaps the most important official report was that prepared by the US Atomic Energy Commission (USAEC). This report – *The nation's energy future*, WASH-1281 – was issued in December 1973, just six months after President Nixon had directed Dr Dixy Lee Ray, then chairman of the USAEC, to 'undertake an immediate review of Federal and private energy research and development activities ...and to recommend an integrated energy research and development program for the Nation'.

In this report Dixie Lee Ray told the President that '1985 is the earliest date by which self-sufficiency can reasonably be expected with this program', which is 'both necessary and sufficient to maximise energy R & D's contribution to the Nation's energy goals' (President Nixon had called for independence by 1980). Another report – *US energy prospects: an engineering viewpoint* – warned that even to achieve independence in a decade would require enormous efforts. This second report, from the National Academy of Engineering (NAE), is particularly interesting because the man who was president of the NAE at the time, Robert Seamans, later became the first head of the Energy Research and Development Administration (ERDA). WASH-1281 said that the President should 'establish an operational Energy Research and Development Administration not later than July, 1974 to plan and coordinate the total program and to direct the major share of the Federal Program'. ERDA did not make it into existence as an operational unit until the beginning of 1975. (This may not have been up to Dixy Lee Ray's timetable, but ERDA's establishment was rapid enough considering that the US political system had been all but paralysed by the 'Watergate' affair until President Ford took over.)

WASH-1281 concentrated on the American situation but it is clearly internationally significant because the US's energy R & D programme swamps all others. The report said that there should be a 'massive concentration of effort on short-term objectives'. It suggested that the federal government should spend $10 billion between 1975 and 1979, and that private industry should match this with $12½ billion.

Another influential American report was the Federal Energy Administration's Project Independence report, which asked two questions relevant to the energy R & D situation: 'What can energy R & D contribute in the short-range covered by the Project Independence Blueprint (to 1985)? What additional R & D ought to be adopted now to help assure acceptable energy options in the time beyond 1985?' On the shorter term the report says: 'The most critical technical problems in the Project Independence period involve increasing oil and gas supplies, using energy more efficiently and using available coal and uranium. Research and development can play a secondary, but necessary, role in overcoming these problems.' The Project Independence report highlighted three areas that are particularly important in the short term: enhanced oil and gas recovery methods, conservation, and greater use of coal. The first of these 'offers a large near term pay off in the terms of increased supply and recoverable resource base if the required research and development, and field testing is accelerated'. Conservation 'will depend primarily on widespread use of existing technology and improved design practices, but R & D aimed at improving consumption technologies and the process of implementing them is needed'. Coal's major problems are environmental, says the Project Independence study.

The report says that: 'The depletion of conventional oil and gas resources dominates the post-1985 period.' One consequence of this is the need for synthetic oil and gas made from coal and oil shale. 'The oil and gas shortfall is so large that major shifts in demand, most probably to electric power, as well as conservation measures will be needed. To achieve these post-1985 goals, new technologies for using energy more efficiently and for shifting to electricity instead of oil and gas must be developed.' The report summarises the long-term energy R & D programme as follows: 'New energy sources not limited by conventional fuel and uranium resources are needed. Technologies, such as the breeder, fusion and solar energy, will take decades to develop and introduce, but it is important to know that we will have one or more of these to support the shift in demand. R & D on these technologies must be pursued as a basis for charting the course of energy development.'

National interests

Other nations could not hope to match the massive energy R & D effort of the US, and it took most of them longer to assess their R & D programmes. Prior to the oil upheavals Britain, for example, had a strictly conventional energy R & D programme with nearly all its eggs in the nuclear basket. A few brave souls had expressed anxiety about the nation's sad neglect of non-nuclear R & D, but 18 months after the oil embargo the chief scientist at the Department of Energy had to confess to a parliamentary committee that it might take him another year to formulate a national energy R & D policy. (This was in February 1975; thus two years were to elapse after the oil crisis before the UK could begin to respond to the energy situation in its R & D policy.) West Germany was quicker off the mark: at the beginning of 1974 the government approved a four-year crash programme costing £250 million. The largest part of this was set aside for R & D on coal gasification and liquefaction, with the next largest share going to coal mining R & D. In October 1974 a Swedish government commission reported on its evaluation of that country's R & D policy. It called for a three-year programme costing Sw.Kr 420 million. The commission recommended that energy R & D should try to get away from the earlier emphasis of energy production. Thus only Sw.Kr 234.9 million were to be allocated to energy production. The rest of the budget was to be split as follows: industrial processes, Sw.Kr 36 million; transport and communications, Sw.Kr 27 million; space heating, Sw.Kr 68.5 million; recovery of energy in goods (by recycling materials or converting garbage into fuel), Sw.Kr 9 million; and general energy systems analysis, Sw.Kr 4 million. The commission proposed that a further Sw.Kr 40.6 million should be set aside for anything that might later prove to be worthy of government support.

Britain, as we have said, responded slowly to the energy crisis. Its major response was organisational rather than financial. At the beginning of 1974 the British government decided that energy was an important enough topic to warrant the establishment of a separate government ministry. Thus the Department of Energy was born. (Britain had maintained a separate Ministry of Power until in the mid-1960s the Labour government had absorbed it into the giant Ministry of Technology.) R & D was the last thing that the infant department was interested in. Britain was hit by energy problems more serious than those of any other nation at the beginning of 1974. Everyone was suffering from oil supply difficulties; on top of this Britain's coal miners were on strike, and the electricity workers were causing trouble. (Both groups of workers had previously proved their power by producing earlier 'mini-crises' by industrial action.) It was not until well into 1974, after a general election changed the government and gave the country a new minister for energy, that the Secretary of State for Energy, Eric Varley, got around to the problem of Britain's energy R & D. His first job was to find a chief scientist for his young department. Britain had always maintained a chief energy scientist on the books, but until 1974 the man whose task it was to advise the government on energy R & D also acted as the chief inspector of nuclear installations. A professional civil servant of many years standing, the nuclear inspector of the day was not disposed to turning energy R & D on its head. And in any case a Select Committee of Members of Parliament had complained about the anomalous position of the chief energy scientist: not only did he safeguard the interests of the public in what was fast becoming a sensitive area, nuclear safety, but he was also advising the government on the type of new energy technology, including nuclear technology, it should be encouraging. With this peg to hang changes on, Eric Varley appointed a new science adviser. He picked Dr Walter Marshall, who was at the time the head of the UK Atomic Energy Authority's Harwell research laboratories. (Marshall continued to do this job when he became chief scientist.) Marshall's appointment was criticised because he was wed to the nuclear camp. As we have seen, most countries, including Britain, had devoted most of their energy R & D effort to nuclear power; it was, therefore, most likely that nuclear scientists were in the best position to under-stand the most significant parts of the energy R & D effort.

Walter Marshall's first job was to take stock of what was already going on. The Department of Energy was responsible for some

Table 3.1. *Britain's energy R&D programme*

	1973–74 (£m)	1974–75 (£m)
Department of Energy direct spending on research and development		
Safety in Mines Research Establishment	1.3	1.1
R&D by outside bodies on behalf of Department of Energy		
Offshore oil and gas	1.945	3.89
For the Nuclear Installations Inspectorate	0.128	0.98
Nuclear R&D (other than by the UKAEA)	12.1	14.1
Other	nil	0.16
Totals	14.173	19.13
Nationalised industries and UKAEA direct spending on energy R&D		
National Coal Board	5.6	6.5
British Gas Corporation	8.0	9.0
Electricity Council and Electricity Boards	25.0	28.5
UKAEA	58.0	70.6
Totals	96.6	114.6

£110 million a year of energy R&D, but most of this was extremely conventional activity in the hands of the energy utilities (see table 3.1). Less mainstream R&D ideas had been all but frozen out by the Central Electricity Generating Board which commanded an inordinate influence over Britain's energy R&D effort. Even some of the less 'way out', and to some people extremely attractive ideas (fluidised-bed combustion of coal was just one) had been all but killed off because the CEGB just was not interested in anything but the conventional approach to energy. The CEGB wanted to burn as much oil as it could lay its hands on; nuclear power was welcome only if it could compete with oil at the current price; coal was something the CEGB would prefer less of (miners were proving to be much too bothersome with their annual industrial disputes). That many energy experts were beginning to get jittery about the oil situation (even before the Middle East war) was not something that the CEGB wanted to know about. Oil was cheap, so it wanted to build oil-fired power stations even if people were saying that the price of oil was bound to go up, and almost certainly by a considerable amount over the life of a power station. When this clearly shortsighted CEGB policy was shown to be misguided, the Board reviewed the various alternative energy resources. As a result of this extensive re-assessment even the CEGB came out in favour of spending some money on new energy R & D projects, on wave power, for example.

The subject of wave power brings up another British report which we really cannot ignore. This is the report, *Energy conservation*, prepared by the Central Policy Review Staff (CPRS) – the British government's 'think tank'. Ostensibly a report on energy conservation – this was, after all, the title of the report – the 'think tank' document turned out to be a somewhat eccentric analysis of all energy options. It was in this report that wave power first obtained any 'official' support. The report recommended that 'The first stage of a full technical and economic appraisal of harnessing wave power for electricity generation should be put into hand.' This was done, and the result was a report from the National Engineering Laboratory. The government went further than this and in November 1974 announced that it was spending £65 000 on a wave power R & D programme.

Perhaps more important than the recommendations on what should and should not be done on the energy R & D front were the CPRS's recommendations on the policy Britain should adopt when establishing R & D priorities. A significant comment was that: 'It is most important that the United Kingdom should not dissipate its scarce resources by doing research and development on all possible ways of energy conservation and substitution.' The report advocated a more international approach to energy R & D: 'Virtually all developed countries are engaging in such work, and we shall be able to profit from the results.' The 'think tank' said that, instead of trying to do everything itself, Britain should accept that 'in many cases the United Kingdom's role is to monitor progress elsewhere'. The Department of Energy accepted this advice and set up an Energy Technology Support Unit as the monitoring institution. ETSU, as the unit soon came to be known, was established at Harwell, thus guaranteeing that many of the energy R & D practitioners outside the establishment would dismiss the unit as yet another arm of the nuclear energy business. (Harwell, after all, was the leading nuclear R & D establishment in the country.)

Energy R & D is unique in the internationalism it has aroused.
While the nations hit hardest by the oil crisis have been seeking,
through international agreements, ways of surviving possible
future oil embargoes, and ways of 'recycling' the excessive money
that flowed into the oil producing nations after they put oil prices
up, they have also been looking for ways in which their energy
R & D might be made more effective. The countries involved did
better in their energy R & D talks than in their other discussions.
One of the more significant moves as far as international cooperation on energy R & D is concerned was the establishment of the
International Energy Agency (IEA). This came out of one of
Dr Henry Kissinger's famous 'initiatives', and was set up under
the auspices of the OECD. The IEA did not come about solely
for the benefit of energy R & D – indeed this part of the package
was way down the list of priority areas – but thanks to the enthusiasm of the people involved, the IEA's first impact was felt
on the R & D front. In January 1975, a little more than two months
after the IEA came into existence, and some time before it
assembled its permanent staff, the R & D sub-group of the IEA
implemented one of the proposals of the initial agreement. This
initial 'Agreement on an International Energy Program' called for
cooperative action on energy R & D. The agreement listed ten
areas of R & D in which the IEA was asked to consider cooperative
programmes 'as a matter of priority'. The agreement says that
long-term cooperation should be considered in the following
areas: conservaton of energy, development of alternative sources
of energy such as domestic oil, coal, natural gas, nuclear energy
and hydroelectric power, and uranium enrichment.

The IEA has no money or research capability. All it can hope
to do is coordinate national programmes and make it very much
easier for different research teams to talk to one another. This is
difficult enough on a national basis, let alone internationally, but
the IEA succeeded in building the framework for greater
collaboration very early on in its discussions. One of the first
things it did was to allocate nine of the ten priority areas to 'lead
agencies'. These were supposed to act as a focus for R & D
planning, bringing together those countries interested in a particular topic to formulate an R & D programme. The nine R & D areas
were allocated as follows: coal technology was put into the hands

of Britain's National Coal Board, solar energy went to Japan's Ministry of International Trade and Industry, radioactive waste management went to the OECD's Nuclear Energy Agency, the European Economic Community was made responsible for both controlled thermonuclear fusion and the production of hydrogen from water, the United States became lead agency on nuclear safety and energy conservation, waste heat utilisation became West Germany's responsibility, and finally, the Netherlands took on the task of overseeing work on municipal and industrial waste utilisation for energy conservation. A tenth topic – overall energy systems analysis and general studies – was not allocated to a lead agency. (This was partly because this was a relatively new topic that had not, at the time, been taken up, to any significant extent, outside university groups. The feeling at the IEA was that energy systems analysis should be left to find its feet before international cooperation stepped in.) Early in 1976 the IEA added seven more R&D topics to its portfolio, including high temperature reactors, geothermal energy, wind power, wave power, and ocean thermal power.

The OECD report on energy R&D contains a concise explanation of why it makes sense to go beyond national boundaries when considering energy R&D:

'Whilst national R&D policies have to be based on the specific conditions applicable to each country, it is because of these differences and the interdependence of all countries that such policies must take into account the international character of the energy problem. Energy raises trans-national problems and countries have to cooperate. Furthermore, cooperation is necessary not merely to avoid unnecessary duplication of effort but also because of the scale of the problems, the rising cost of research and the complexity of the technologies involved.'

Time for thought

The thought that has gone into energy R&D policy formulation may have delayed the start of work on some projects, but as long as the delay is not too long the time is far from wasted. It is a sign of the more hard-headed approach to energy R&D. More and more voices have come out against any policy that smacks of wishful thinking. While the longer-term energy systems are much talked about, few people would advocate that we should mark time with the more down-to-earth systems and hope for something to show up and snatch us from the jaws of energy famine at the last minute. But we must beware of placing excessive faith in the more esoteric energy systems that must inevitably take up much of the space in a book devoted to energy R & D and longer-term prospects. It is worth taking note of a warning spelled out by Chauncey Starr in 1972:

'The development of new speculative energy resources is an investment for the future, not a means of remedying the problems of today. Unfortunately, many of these as yet uncertain and un-developed sources of energy are often misleadingly cited publicly as having great promise for solving our present difficulties. In addition to their technical uncertainties, many of these speculative sources are likely to be limited in their contribution, even if suc-cessful. The attraction of "jam" tomorrow may persuade us to neglect the need for "bread and butter" today.'

In the following chapters we will look at the bread and butter as well as the jam. Coal and oil must continue to be our 'bread' for some time and it is encouraging to see that both are to be more closely pursued as R & D topics. Nuclear power can only be described as the 'butter' that we now spread on top of the bread. As far as nuclear R & D goes, it is doubtful if even more spending is justified. It already takes the lion's share of national R & D budgets. There are, however, some significant doubts about the way in which nuclear priorities are set. There are, for example, four different programmes under way in pursuit of one particular nuclear option (the liquid-metal-cooled fast breeder reactor) while it just could be that we are ignoring an alternative that might meet our requirements more easily (such as the gas-cooled breeder reactor). Energy's 'jam' comes in many flavours. This is, as we shall see, sometimes a matter for taste and sometimes the only option.

Chapter 4

Coal

Coal sparked off the industrial revolution and set us on the exponential curve of rising energy consumption. There are those who
see the world's massive coal reserves as a reason for returning
to coal as the number-one energy source. There may be massive
coal reserves, as we saw in chapter 2, but it has to be recoverable
if it is to be of any use. Some of the remaining coal could be
mined easily enough. In the US large quantities of coal lie in thick
seams very near the surface. The way to this coal is through
strip mining, tearing the layer of soil from on top of the coal
and digging up the coal with huge mechanical diggers. This
approach has its environmental costs, hence there is strong
opposition to the development of strip mining in the US, where
people are now keenly aware of their environmental responsibilities.

Even if the world could dig up all its energy requirements as
coal, we would find it difficult to use the energy. As we saw earlier
the way in which we use energy dictates the form of the energy
carrier. Coal is little use for transport, for example; it comes in
large heavy lumps while cars and aircraft rely on liquid fuels to
keep them mobile. (We have had coal-fired steam trains, but the
railways have been seduced by the convenience and superior
performance of diesel engines.) Coal will make a comeback as a
transport fuel only via electrification, or liquefaction of coal. The
energy system of the developed world depends mostly upon liquid
fuels, gaseous fuels, and electricity generated from various fuels.
Coal will take over from oil and gas, and halt the flow of nuclear
electricity, only if it can fill their roles in some way and at a price
that makes it an attractive substitute. In theory coal can do everything; but theory has to be taken out of the laboratory into industry.

Before coal can regain any of its former glory, considerable
research effort has to be expended. The OECD study of energy
R & D summarises the situation:

'A big increase in coal production necessitates the development of cheaper and more efficient methods of coal mining. Moreover, to remove the polluting components of coal and convert coal into more versatile liquid or gaseous fuels, substantial R&D efforts are necessary. Even when these projects are based on known principles or processes, three, five or even ten years of effort will be needed to develop these processes to the point of commercial application.'

Despite the impression that the coal R&D community may try to give, their subject has not been ignored. The countries with appreciable coal reserves have spent considerable sums on coal R&D. In the 1973/74 financial year Britain, for example, spent £5.6 million on coal R&D through the National Coal Board. In the US support for coal R&D has soared over the past few years. In the fiscal year 1974/75 the US spent nearly $180 million on coal R&D; and the expected expenditure for the following year was $100 million more than this. Most of Britain's spending was on new mining technology; in the US about a fifth of the budget went to this area. The OECD report pointed out that without adequate supplies of coal, all the clever utilisation techniques are worthless, therefore coal mining R&D should receive priority support. It is, therefore, a short-term priority with long-term implications. To a certain extent there is plenty of scope for improving mining technology without introducing new technology. In Britain, for example, existing technology can improve the viability of the coal mining industry. Because the country's coal industry was almost written off before the oil embargo and the subsequent price rises, the government had been reluctant to invest money in new mining technology. Once coal became a competitive fuel again the situation changed and a new investment programme was set under way.

In the long term it may be possible to dig coal out from underground coal faces without having miners go down to do the job. Dr Meredith Thring is something of a self-confessed fanatic on this topic. He believes that much of our mining R&D effort should be devoted to the development of unmanned robots that could dig out the coal under control from the surface. While there are many people in the coal industry who accept that automatic coal mining would be an admirable innovation, past experience with automation has tempered the enthusiasm for grandiose projects. Britain's coal industry has come up against problems of equipment reliability when putting new machinery into the hostile working environment of a coal mine. The British experience shows that it pays to proceed with caution, and that it is unwise to jump too far too fast. When the National Coal Board tried to introduce massive automation at its Bevercotes colliery, it found that the technology just was not advanced enough; in particular, it was difficult to keep the machinery operating in the appalling working conditions. The NCB now plans a more orderly move towards automation. Gradual automation will have an impact on all coal mining operations, from the coal face onwards, and will steadily reduce the number of people at the coal face and improve the efficiency of those miners who have to operate and oversee the machines.

One way in which the amount of coal extracted from an underground mine could be increased would be to organise the mining differently. The so-called 'room and pillar' technique leaves behind as much as two-thirds of the coal in a seam as large pillars of coal acting as roof supports. An alternative way of mining known as 'longwall' mining allows 60 to 80 per cent of the coal in a seam to be cut out; it can be done along a face several thousand feet wide. Mechanical roof supports stop the seam from collapsing. As the mining proceeds into the face, the roof supports would move forward, allowing the roof to collapse behind them. The skill lies in understanding the soil mechanics so well that the collapse is controlled. According to Edmund Nephew, of the Oak Ridge National Laboratory in the US: 'The longwall system of mining offers important advantages which make it highly attractive. These include high productivity, high resource recovery, and improved mine safety conditions. The face equipment can be operated by a small crew of from eight to ten men, and production of up to several thousand tons of coal per day can be achieved.'

Productivity per man is what new mining technology is ultimately all about. New machines will eventually be needed to extract coal from seams presently unworkable; but machines are now mostly used to boost productivity rather than to exploit difficult seams. Between 1947 and 1973 Britain's National Coal Board suffered a fall in coal output from 184 to 127 million tons

▶ When the National Coal Board wanted to improve Britain's coal mining R&D programme it decided to concentrate on remote control and automation of the equipment used in the severe environment of the coal mine. An NCB report said: 'Unlike previous innovations remote control or automation can be applied to every phase of the mining operation, and for this reason its potential impact is the greater.'

Underground mining is a complex series of operations. Coal has to be cut from the face, loaded on to a transport system, and then carried away from the face to the mine surface where it is sorted and prepared for the customer. While this is going on the mine roof has to be shored up, men have to be carried to and from the coal face, and the machinery that does all this has to be maintained in working order.

Remote control and automation will have several effects on mining operations. Sophisticated mining machinery is expensive, hence the improved efficiency that should come with remote control and automation will be a significant gain; as will the better use of mine manpower, another commodity that is becoming expensive and scarce.

a year; but during that same period the output per man-year rose from 262 to 473 tons, and the output per man-shift from 3 to 7 tons for underground workers. Edmund Nephew quotes figures for the US of productivity rises from 5 tons per man-day in the 1940s to 20 tons per man-day in the 1970s. Longwall mining, he says, could give productivities approaching 509 tons per shift for each man on a 10-man face crew.

Much of the improvement in productivity in the US has been through the growth of open-cast mining (also known as strip mining) which now provides about a half of the coal dug up in the US and can recover 80 to 90 per cent of the coal. One mine in Wyoming has achieved an astonishing 80 000 tons a year from each of its workers. Here the technical challenge does not lie in finding ways of digging up the coal – all you need is bigger earth-moving machines – but in alleviating the environmental problems.

Strip mining can leave large scars on the landscape, as trees and topsoil are pushed aside to let the machines get at the coal. Much of the coal extracted in this way contrains a high percentage of sulphur. When this coal and the debris left after mining it are exposed to the air and rain, sulphuric acid washes off into local water. In parts of the US many thousands of miles of streams and rivers have been contaminated by acid water from both open-cast and deep mining. Add to this the large areas of land that have been devastated and turned into a useless wasteland by strip mining and you begin to see why American environmentalists are fighting so hard for controls on the strip miners. On the other hand, the coal industry complains that it will have to abandon some of the richer coal deposits if the environmentalists have their way.

The only way round this dilemma is to ensure that only those areas that can be reclaimed are strip mined. But someone still has to say what constitutes reclamation. This is far from an academic point. Reclamation can be done to varying degrees, at costs that can go as high as $5000 an acre. If strip mining is delayed until satisfactory reclamation techniques are developed, coal production will certainly be below what the industry could achieve without any environmental constraints. Such brakes on the despoilers could also act as an incentive to the development of new reclamation techniques. These can be quite simple – in some cases it is enough to reorganise the mining to minimise the area of land that is

▶ 'I might have known it, Albert Scrimthorpe,
"Another hard day down t' pit" indeed!'

disturbed by moving freshly disturbed overburden to reclaim previously mined land. As far as strip mining is concerned, the coal industry does not really have to develop new coal mining technologies as much as it has to convince people that it does not have to leave the land looking like a lunar landscape when it has removed the coal and moved on to a new site. Thus the viability of the most rapidly growing sector of the American coal mining industry depends on research into ways of reclaiming land rather than ways of ripping coal out of the ground.

New uses of coal

After a time in the doldrums coal utilisation research is something of a glamour business again. Just a few years ago Britain's coal scientists were having trouble convincing the government and some of the people in the coal industry that it was sensible to spend money in pursuit of new ways of using coal. The anti-research argument was that more than half the country's coal was consumed in power stations, and very little of the proposed coal utilisation research would benefit the industry's existing customers who should not, therefore, be expected to pay for the R&D. Fortunately, at that time it was possible to make significant progress in the laboratory without having to spend money on large pilot plants. Work went on successfully at this 'fundamental' level. The high quality of the work maintained Britain's prestige in this area. So much so that when the US wanted to increase its spending on coal R&D it started paying for work to be done in the UK. And when the International Energy Agency (see chapter 3) wanted to nominate a 'lead agency' to coordinate coal R&D in different countries it turned to the National Coal Board, which took on this task.

Coal utilisation research aims to put coal in a position to take over certain energy markets from oil and gas. There are numerous ways of going about this. One approach is to burn coal in power stations where oil and gas have been used in the past. (This means restoring the economic advantage of coal over these relatively new power station fuels.) A second approach is to take coal and to process it to produce substitutes for oil and gas. If coal can be liquefied at the right price on a large scale – and South Africa has

▶ Today's coal-fired power station, such as the one at Ratcliffe-on-Soar near Nottingham, is a huge affair. In Britain these giant power stations are built with tall chimney stacks which help reduce pollution near the power station.

In the US there has been a violent debate as to the best method for cleaning-up power station stack emissions. Coal is not the cleanest of fuels because it often contains sulphur which produces sulphur dioxide when burnt.

There are two approaches to the sulphur dioxide problem: either you eliminate sulphur from the exhaust gases, or you send the exhaust gases into the air through tall chimney stacks which create a tall plume that carries any pollutants high into the air where they are diluted and dispersed. Britain has adopted the 'tall stack' approach; but there are complaints that this merely dumps the pollution on someone else. (The Scandinavians object to Britain's sending them 'acid rain'.) In the US there is strong pressure to fit 'scrubbers' to remove sulphur dioxide from power station emissions. There are, however, industrial complaints that scrubbers are not yet ready to take on the job – the environmentalists claim that this is an excuse for not investing in the equipment needed to help clean up the environment.

been making oil from coal on a commercial scale for a quarter of a century – then it stands a chance of replacing oil as a transport fuel. This sector of the energy consuming business depends on the nature of its fuel more than any other.

Gasification and liquefaction of coal are not yet crucial to the continued supply of oil and gas; but when these two fuels begin to run out coal could be the most attractive alternative energy source. And if the price of oil from the Middle East is high enough oil and gas made from coal could be cheaper to countries with abundant and inexpensive coal. The US's desires to reduce foreign spending on imported oil and to free the country from the risk of blackmail by other countries have added to the attraction of artificial fuels even though natural gas and oil are far from running out in other countries. It is also worth noting that the oil-producing countries have at various times said they will fix the price of their oil at the level determined by the alternatives to their oil. This means that crude oil prices will be pegged to the price of oil derived from coal. In theory this sounds like a reasonable way of setting the price of crude; but the history of oil prices shows that such 'rational' pricing policies always break down in the face of political pressures. This is why the companies interested in establishing large and expensive coal processing plants want some guarantee that if the price of crude oil falls and makes synthetic oil and gas an unprofitable proposition they will be protected in some way.

Why go to the trouble and expense of turning coal into gas or oil if you want to generate electricity? There are reasons for going through this process, as we shall see later, but there is at least one technology available that could improve electricity generation without going this far. Oil has been burned in power stations because it was much cheaper until very recently; it also has a slight environmental edge over coal in that any sulphur in oil can be more easily removed than sulphur in coal. The pollution problem is so severe that in the US the attraction of inexpensive coal is blunted by environmental constraints. Cheap coal often contains a lot of sulphur, leading to high sulphur dioxide concentrations in power station stack gases. The simplest way round this difficulty might appear to be to clean up the combustion gases before they are released to the atmosphere. Unfortunately, this technology is not very popular with some electrical utilities, who maintain that 'scrubber' technology is unproven, expensive, and unreliable. Britain relies on tall chimneys to 'solve' its pollution problem. The hot combustion gases are sent up tall chimney stacks so that the pollutants rise to a very great height and are dispersed. This is hardly a genuine answer to this problem, even if the environment can disperse and absorb the sulphur dioxide. (The Swedish government, for one, *knows* that its rain is polluted by Britain's sulphur dioxide.) Dispersal is not a permanent solution. And the OECD energy R & D report said: 'For the next few years, R & D to develop clean coal-burning technologies will yield quicker results than research on any other coal problem with the exception of mining. Thus, clean coal-burning R & D should certainly receive massive support.'

We can start very near the beginning of the electricity generating process if we want to clean up coal-fired power stations. Fluidised-bed combustion is a possible answer to the pollution problem, and to the problem of burning poor quality coal. The fluidised bed is not a new idea, but the time now seems ripe for its development. In a fluidised bed, coal is crushed to form particles roughly $\frac{1}{8}$ to $\frac{1}{4}$ inch in diameter. When air is blown through a bed of these coal particles at the correct pressure the particles will 'float' on a cushion of air. The pressure has to be just right so that the particles are 'fluidised' without being blown out of the bed. A fluidised bed could be the basis of a direct combustion process, or a gasification system. If the bed is hot enough, the flow of air through the bed can lead to almost total combustion of the coal.

The heat generated in a fluidised bed can be extracted by immersing steam pipes in the bed so that the steam can drive a conventional steam turbine. Alternatively, if the fluidised bed operates at a raised pressure, of perhaps 10 to 15 atmospheres, steam tubes can extract some of the energy and at the same time the hot exhaust gases coming from the combustor can carry energy to a gas turbine. As the gas passes through the turbine its pressure drops to atmospheric pressure, turning the turbine as it does so. Even when the hot gases are at atmospheric pressure further energy can be extracted by recovering the heat energy as hot water for the steam turbines.

The pressurised fluid bed can extract and put to use significantly more of the coal combustion energy. Today's coal-fired plant, which burns powdered coal by 'spraying' it into a combustion chamber with hot air, has a maximum efficiency of 40 per cent or so. This means that 60 per cent of the energy in the coal is lost as hot water and warm air. Fluidised-bed combustion might have an efficiency up to 45 per cent. This apparently small increase of 5 per cent in combustion efficiency actually yields 10 per cent more *useful* energy from the coal that goes into a power station. One side-effect of this improved efficiency is a reduction in the thermal pollution of rivers by power station cooling water.

Fluidised beds have another environmental advantage: they can be designed in such a way that they pollute the air less than conventional burners. In particular, the high sulphur coals that are most unpopular among electricity utilities could be burned in a fluidised-bed burner with minimal pollution. Limestone or dolomite, natural carbonates, can be added to a fluidised bed to react chemically with the sulphur dioxide produced by the combustion of the sulphur in the coal. This produces a solid sulphate that can be disposed of or recycled. (Disposal of the solid sulphate could produce another environmental problem in the same way that stack gas 'scrubbers' produce a wet, sulphate-laden material that has to be got rid of.)

Britain was an early starter in the development of fluid-bed combusion; but the NCB's R&D department was trying to sell the idea when oil, gas, and nuclear power were threatening to oust coal from power stations other than those that had already been built. The idea languished in the UK until the country's coal industry made a comeback following the rise in oil prices. The NCB now plans to build a large prototype pressurised fluid bed. During the ten years or so since the NCB first tried to arouse interest in the UK, the US has picked up the idea of the fluidised bed and has set up several R&D programmes. An American estimate of the cost of developing a pressurised bed put the bill at $75 million. And it might take eight years to turn the system into a commercial product. The time and money would be spent proving that fluidised combustion, which has so far been tried out on only a small scale, will work when bigger beds are built. The problems that have to be researched include the wear of components. (For example, can heat-exchange tubes withstand the continuous

abrasion they will experience in a bed of small rough particles?)
And before a pressurised bed can be put in front of a gas turbine,
the small particles that are inevitably carried off in the hot combustion gases will have to be removed; if not they will find their
way into the turbine and damage the turbine blades.

As well as its efficiency and environmental superiority over
rival coal burners, the fluidised bed can burn poorer quality coal
containing more 'ash' than is normally acceptable. The pressurised
bed also makes better use of engineering materials. An atmospheric
fluid bed produces ten times more heat than is produced in the
same volume in a pulverised-coal burner. The fluidised bed will
have better heat transfer from the combustion to the steam
circuits; the temperature of the fluid bed is significantly lower
than in a conventional burner. (The fuel and air are in better
contact, so the coal burns at a lower temperature.) This means
that the heat-exchanger surfaces can be put into the fluid bed.
And because the burnt fuel does not get so hot, it shows less
tendency to stick together in large lumps, making it easier to
operate the combustor. So a fluidised-bed combustion system
would be smaller than a conventional burner, and would last
significantly longer because of the less severe combustion conditions.

The next stage of fluidised-bed development, after a simple
burner, would enter the realm of coal gasification and liquefaction.
Various ideas have been put forward for using a fluid bed as part
of a processing plant for turning coal into liquid fuels, or into gas
that would have an energy content high enough to make it suitable
for long distance transmission as a substitute natural gas (SNG).
To be useful as a pipeline gas, any gas made from coal must contain enough energy. Poorer quality gas is not suitable for transmission over any distance because more gas has to be shifted to
carry the same amount of energy, adding considerably to the cost
of gas transmission. Alternatively, a poorer quality gas can be
produced for immediate combustion in a gas turbine. This may seem
to complicate the system when pressurised fluid-bed combustion
can do the same job without the intermediate gas-making and combustion stages, but the extra step would give a further boost to
the efficiency of coal utilisation. A coal gasifying fluid bed could
achieve an efficiency as high as 50 per cent in turning the energy

of coal into useful electrical energy. Some preliminary estimates have suggested that a gasification plant might not cost any more than a straightforward combustion plant. In both cases the cost of energy from the power station might be 10 per cent below the cost of energy from a conventional coal-fired power station.

Gasification and liquefaction

There is no shortage of ideas as to how coal might be turned into liquid or gaseous fuels. Few of them have been thoroughly researched. However coal gasification and liquefaction have been carried out on a commercial scale. The British gas industry once almost completely depended on gas made from coal, but Britain gave up its artificial gas industry when natural gas started to come ashore from the North Sea. (Coal lost its place in the country's gas works even earlier, when oil became a cheaper feedstock.) For 25 years or so South Africa has been making a synthetic oil from coal. The coal liquefaction technology used by SASOL, the state-owned South African oil-from-coal company, is perfectly satisfactory when coal is cheap; it is, however, less efficient than some of the newer techniques now under development.

Coal gasification technology now centres on attempts to develop techniques that can produce a substitute natural gas. Britain's gas works produced a lower quality gas, containing about half the energy of natural gas which is mostly methane.

The aim of all coal gasification and liquefaction technology is to alter the ratio of hydrogen to carbon of the coal feedstock, which involves adding hydrogen to make the hydrogen/carbon ratio more like that of oil or natural gas. The weight of the carbon in coal is about 14 to 30 times the weight of the hydrogen. Coal consists of large carbon-rich molecules which have to be broken down into smaller molecules if the coal is to yield liquids or gases; natural gas for example is mostly methane which has a carbon/hydrogen mass ratio of three. Hence complete gasification of coal requires the addition of large amounts of hydrogen. The hydrogen can be made by reacting steam with hot coal in a series of reactions that add up to the overall reaction:

$$2C + 2H_2O = CH_4 + CO_2$$

An alternative to adding hydrogen to coal is to remove some of the carbon. Pyrolysis is one way of achieving this. Pyrolysis is no more than the process of heating coal without letting any air or steam into the system. The reaction can be altered to change the product yield, generating a mixture of gas liquid and char (the carbonaceous solid residue left over when the hydrocarbons have been removed). Pyrolysis is conceptually simple and not very difficult to put into practice. It is however an inefficient way of converting coal's chemical energy into gaseous energy. Only a quarter of the energy present in the coal feedstock is retained in the gas produced.

Various pyrolysis projects are under way. Most of them are dedicated to the development of coal liquefaction techniques rather than a viable gasification technology. The COGAS SNG gasification process is being developed by both the COGAS Development Corporation and the National Coal Board. (The NCB is working under a COGAS project.) Coal liquefaction projects often have a pyrolysis stage. The US Office of Coal Research has funded several projects including the COED (Coal Oil Energy Development) process which the Office of Coal Research (OCR) described as 'the most advanced coal-to-clean liquid fuel conversion technique at this time', in its 1973–74 annual report.

The COED process which produces synthetic crude oil, involves pyrolysis in four fluidised beds at progressively higher temperatures. This produces a liquid feedstock which is subsequently treated with hydrogen to upgrade it and to reduce sulphur, nitrogen, and oxygen to a low level in the synthetic crude oil. In 1974 the project was budgeted at more than $20 million. The test plant processed up to 36 tonnes of coal a day in a series of test runs that started in 1970. As with all coal processing techniques, the COED process will have to cope with assorted types of coal as its feedstock. The development programme of any coal conversion technique inevitably involves processing various batches of different varieties of coal. The COED pilot plant has produced 1 to $1\frac{1}{2}$ barrels of oil from each tonne of coal feedstock.

The NCB has described a pyrolysis unit for making synthetic natural gas and producing electricity. Here the pyrolysis stage also involves introducing hydrogen into the powdered coal. The pyrolysis-hydrogenator yields a gas that is processed to remove sulphur dioxide (generated as a result of the sulphur in the coal feedstock) and boost the methane content. The char that is also produced at this stage is the feedstock for a fluidised-bed gasifier which also has steam and oxygen fed into it. The reaction in this stage generates hydrogen that is passed into the pyrolysis-hydrogenator unit. The fluidised-bed gasifier yields a carbon-rich residue, and this goes on to yet another fluidised bed (a pressurised combustor). Here the carbon burns in air and the heat raises steam for electricity generation.

The various coal gasification techniques produce gases of different qualities. The gases have varying energy contents. (The energy content is usually expressed in British thermal units, Btu's, per cubic foot.) The aim of most coal gasification processes is to generate a 'pipeline quality' gas that is interchangeable with natural gas, which means making a gas with 950–1050 Btu/cu.ft. Some gasification processes make a gas with about half this energy content (low-Btu gas). This gas is uneconomical for long distance transmission – you have to shift twice as much gas to move a given quantity of energy – but the gas is valuable if it can be made and consumed on the same site. An obvious candidate for low-Btu gas production is a power station where the resultant gas would be burned in gas turbines to generate electricity. Gasification can reduce the pollution that goes with burning coal. And the efficiency with which the energy in the coal is turned into electricity can be significantly higher for a coal gasifier than for a combustor. Hence this more complex technology could be viable if the energy savings are not grossly outweighed by higher costs of building the coal processing plant.

Pipeline quality gas, or some other natural gas substitute, will be essential if the world's gas distribution and transmission systems are not to be left idle when natural gas runs out – hence the considerable interest in coal gasification in the US, where the prospect of natural gas shortages is more immediate. In the US the OCR has said that its objective is to see 'at least one economically sound gasification process available in the United States by 1980'. In the 1975 fiscal year the US spent more than $57 million on high-Btu gasification, and more than $22 million on low-Btu gasification. And the OECD's energy R&D report lists more than 20 gasification projects.

The American coal gasification programme is centred on four possible processes (each with its own pilot plant in operation): HYGAS; CO_2 Acceptor; BI-GAS; and Synthane. All four plants make a gas that is a mixture of hydrogen, carbon monoxide, and as much methane as possible. The different plants achieve the same basic chemical reactions in entirely different ways. According to the US Office of Coal Research 'the HYGAS process is the most advanced of the coal-to-high-Btu gas schemes under development'. HYGAS is a corruption of 'Hydrogasification', and the project is managed by the Institute of Gas Technology, with a total anticipated budget of over $30 million (at 1974 prices). As the name suggests, HYGAS needs extra hydrogen to drive the process. (Hydrogen production is one of the projects associated with the HYGAS programme.) The most promising hydrogen production technique is steam–iron gasification, in which hydrogen is made by steam oxidation of iron – the iron oxidises as the steam decomposes, yielding hydrogen gas. A second technique, steam–oxygen gasification, has the advantage that it consumes char from the HYGAS plant. Steam and oxygen react with this char to yield a hydrogen-rich gas that goes on to the HYGAS reactor. This is a two-stage plant. In the first reactor stage about 20 per cent of the coal is converted to methane – ungasified solids drop down into the second reactor stage. This second step is a fluidised-bed reactor where an additional 25 per cent of the coal feedstock is turned into methane. The organic carbon material remaining after this stage is available for combustion to generate heat, or for the production of hydrogen. The HYGAS pilot plant was designed to convert 75 tons of coal a day into 1.5 million cubic feet of high-Btu gas.

The BI-GAS process may be less advanced – the pilot plant was due to start working in 1975 – but it has some advantages over the HYGAS technique. According to the OCR, 'coal fed into the BI-GAS gasifier does not require pre-treatment; and the gasifier itself appears to have the potential for high volume yields – more gas per unit of gasifier volume than any other process now in the pilot plant stage'. The BI-GAS pilot plant is not big, converting 5 tons of coal an hour into 100000 cubic feet of pipeline quality gas. A commercial plant would produce as much as 250 million cubic feet of gas a day from 12000 tons of coal. Gasification also takes place in two stages in the BI-GAS plant – stage two sits atop stage one. Coal and steam are fed into the top of stage two; oxygen and steam, together with char carried out of the plant in the product gases, are fed in at the bottom of stage one. Hot synthesis gas is generated in stage one. This rises up through stage two, where it is converted into a methane-rich gas by further reactions with the coal and steam. The BI-GAS pilot plant will cost around $35 million.

The CO_2 Acceptor process takes much of its energy from a chemical reaction between limestone and coal. In this process coal is ground, dried, and fed into a gasifier where it is heated in the presence of steam to a temperature of 800 °C. (The pressure in this stage is 1000–2000 kN/sq.m.) Crushed limestone preheated to 1000 °C is fed into the top of the gasifier. The limestone particles filter down through the gasifier. As they do so they take part in a chemical reaction, the net effect of which is that the limestone absorbs carbon dioxide, generating heat in the process. Spent limestone and carbon residue go on to a regenerator where the burning carbon generates heat which regenerates the limestone by driving off the carbon dioxide so that the limestone can be returned to the gasifier. The ingredients of the pipeline gas are released by the heat and chemical reactions between steam and coal in the gasifier. A final methanation stage turns the gas from the gasifier into pipeline quality gas. A 40 ton per day pilot plant in Rapid City, South Dakota, is evaluating the CO_2 Acceptor process in Rapid City. The 1973–74 OCR annual report put the cost of the pilot plant at $22 million.

The fourth coal gasification technique is the Synthane process. Coal is gasified in a fluidised bed to a gas containing methane, carbon monoxide, hydrogen, carbon dioxide, water vapour, and impurities. Steam and oxygen are fed into the gasifier along with the coal. The gasifier operates at a pressure of 7000 kN/sq.m and temperatures up to 1000 °C. Synthesis gas comes out of the top of the gasifier while unconverted coal or char comes out of the bottom. The char can be burned to provide the heat needed in the process. After it has been cleaned up, the product gas goes on to a methanator where carbon monoxide and hydrogen are converted to methane. (This step boosts the heat content of the gas from 500 to over 900 Btu/cu.ft.)

The various coal gasification techniques ultimately depend upon the same chemical reactions, but they differ in that they can handle different types of coal. The research programmes now under way are evaluating the engineering aspects of the rival plants. The cost of the American R & D effort is high and rising.

In the fiscal year 1975 the US Office of Coal Research spent some $40 million on high-Btu gasification, and a further $50 million on low-Btu gasification. And about the same amount, split differently, was asked for in the following year. The total cost of the four pilot plants is put at around $140 million. But this is only the beginning. In 1974 the National Gas Survey of the Federal Power Commission issued a study on 'future gas supplies from alternate sources'. This gave some idea of the scale of the industrial development that would be needed to bring commercial gasification plants on stream:

'A typical coal gasification plant with a nominal capacity of 250 MMcu.ft/day operating at 1000 lbs/sq.in and an average thermal efficiency of 60 per cent, will require an annual bituminous coal feedstock of six million tons. Capital requirements (in 1970 dollars) for such a plant have been estimated between $260 million and $305 million for a deep mined eastern site, and between $175 million and $230 million for a surface mined western site.' The requirement for 6 million tons of coal a year is not that straightforward, according to the report, which said that 'during 1971, only one mine produced over six million tons of subbituminous coal'. Other constraints will be the availability of water and the difficulty of finding the money to build the gasification plants. The money will not be forthcoming from the traditional sources until coal gasification has been proved to be an economically feasible operation. Thus the US government will probably have to fund this work for some time yet, at least until the next generation of gasification plants is built and operating.

The size of the R & D effort that will be required to bring about a coal-based SNG industry was outlined in a report prepared by the American Gas Association (AGA). This report (*Gas industry research plan 1974–2000*) said that 'during the next three decades, the overall program for necessary research and development represents expenditures of $2.65 billion of which $1.32 billion will be spent in the 1974–1978 time period. This expenditure is justified to ensure maximum supply and lowest cost gas to the public.' The industry that might grow out of this R&D effort could be huge, as was suggested in the National Gas Survey report on future gas supplies from alternate sources. This said that by 1990, the coal gasification industry may have the dimensions shown in the following list:

Number of gasification plants	15–63
Plant capacity	1.5–5.1 trillion cu.ft
Cumulative production	8.2–26.2 trillion cu.ft
Feedstock requirement	129–537 megatons/year
Cumulative feedstock consumed	852–2757 megatons
Cumulative investment in mines	$0.9–3.5 billion
Cumulative investment for plants	$3.9–15.8 billion
(Total cumulative investment	$4.8–19.3 billion)

Artificial oil

Coal liquefaction technology is some way behind gasification technology. However, as with coal gasification, a commercial coal liquefaction system has been in operation for some time. This is the SASOL plant in South Africa, which grew out of the technology developed in Germany during the Second World War. As with the 'proven' gasification technique, the SASOL system could not be developed into a viable commercial system that could take a large share of the energy market. The low efficiency of the coal-to-liquid conversion is the stumbling block. In the SASOL plant coal is gasified, and the resultant synthesis gas is cleaned before flowing on to a liquefaction plant. Gasification accounts for nearly two-thirds of the capital cost of a SASOL liquefaction plant, which is based upon the Fischer–Tropsch technique. This system is less efficient than some of the newer liquefaction techniques. In South Africa the price of coal is low, hence the SASOL process is an attractive way of reducing the country's spending on foreign oil. In fact, the higher price of oil that came into effect in 1974 prompted SASOL to go ahead with a new liquefaction plant ten times bigger than the existing complex. In countries where coal costs are higher than South Africa's, the synthetic crude from a Fischer–Tropsch liquefaction plant would be uncompetitive. One American estimate put the cost of synthetic crude from this process at $12 a barrel.

▶ 'In this experiment we are working on the
conversion of coal into whisky.'

The existing SASOL plant is far more than a simple synthetic crude factory. A sizeable chemical complex is attached to the plant, which produces something like a third of South Africa's nitrogenous fertiliser, among other things. (This makes use of the vast quantities of nitrogen that are a by-product of oxygen generation for the coal liquefaction plant.) The SASOL complex has a daily intake of 15 000 tons of air, 8 million gallons of water, and 7000 tons of coal. Each ton of coal yields about $1\frac{1}{4}$ barrels of oil.

The low efficiency of the Fischer–Tropsch liquefaction system accounts for the high price of the oil it produces. If $12 a barrel really is the price of this oil, then the technique will not make large inroads into the oil market. While rival liquefaction techniques are less developed, it seems that they are already capable of making oil significantly below this price. The OECD energy R&D study says that 'for $5 to $7 per barrel oil, coal conversion technologies could, probably not at the beginning but in the long run, become economically attractive, certainly with United States coal prices. For $7 per barrel many known and projected conversion technologies are competitive and could produce quickly increasing quantities of synthetic fuels during the late 1980s, again based on United States coal prices.'

We have already looked at the COED pyrolysis approach to coal liquefaction. Several countries are also working on solvent refined coal (SRC), with the American programme inevitably much larger than any other. In the SRC process coal is dissolved in a heavy solvent under a moderate hydrogen pressure. The resultant liquid is filtered to remove ash and any undissolved organic material. The final product has a melting point of 175 °C. At first the US Office of Coal Research supported a programme to produce a solid fuel from coal, but subsequent studies showed that the material can be hydrogenated to yield a whole range of liquids. The OCR has spent nearly $20 million on a pilot plant which can process 50 tons of coal a day to produce 30 tons of SRC. In 1973 the OCR estimated that it would cost $75 million to build a commercial plant that could process 10 000 tons of coal a day (3.3 million tons a year). A basic SRC plant that does not process the product through further hydrogenation stages might be one way of burning coal in power stations without polluting the environment.

The SRC process lies somewhere between the two extremes of

coal processing: pyrolysis (as in the COED process), and complete gasification followed by liquefaction (as in the SASOL plant). Pyrolysis liquid yields are low, leaving behind large amounts of solid char. Gasification wastes energy. Other liquefaction techniques are less developed than these three, which form the basis of many of the alternatives. Exxon, for example, is working on a solvent/hydrogen process, involving a catalytic reaction stage outside the liquefaction stage. Another route to coal liquefaction is to make methanol, which can be produced in variants of SNG technology. ICI's methanol synthesis from synthesis gas produced by steam-reforming light hydrocarbon fuels already employs the requisite technology.

The oil companies are showing considerable interest in these new coal technologies. (Their interest in coal is revealed by their gradual take over of much of the American coal industry.) The Gulf Oil Corporation is working on a 'catalytic coal liquefaction' (CCL) project. Bench-scale CCL experiments have been followed up by the construction of a pilot plant that can convert a ton of coal a day into three barrels of liquid fuel. This pilot project cost $1½ million to build. Gulf estimates that a commercial plant taking in 33 000 tons of coal a day could cost $600 million, and would yield 100 000 barrels of liquid fuel a day.

It is not clear when coal liquefaction and gasification plants might be ready for commercial development. To a certain extent it depends upon the price of oil, and on the urgency with which the US wants to make itself independent of imported oil. If this second factor ever became the overriding consideration the US could build SASOL-type plants immediately. And it could turn to the Lurgi-type plants that the British Gas Corporation has operated to make towns gas if it wanted to build an instant synthetic gas industry. British Gas is operating a Lurgi gasification plant as a test bed. This plant, at Westfield in Scotland, has already won R & D contracts worth more than $10 million from the US, where several Lurgi gasification plants are being built. For example, at Four Corners, New Mexico, the El Paso Natural Gas Company is building a coal gasification complex with a coal consumption of 26 600 tons per day. This development was first scheduled to start up in 1976/77 at a cost of $400 million (including $65 million for coal mining capacity).

In the longer term coal processing will probably develop on multi-purpose sites. Coal will be turned into oil and gas, as well as other chemical products, at a single conversion complex. The National Coal Board refers to such a concept as a Coalplex: the Office of Coal Research prefers the term Coal–Oil–Gas (COG) refinery. A dual purpose complex makes sense because the techniques needed to turn coal into oil or gas are similar, and any single-function plant inevitably produces both gaseous and liquid fuels. A study of a COG refinery, carried out in the US by Chem Systems Inc. in 1971, looked at the prospects for a refinery that would consume 57 000 tons of coal a day, along with 7740 tons of oxygen. These raw materials would yield 1800 tons of sulphur, 7660 tons of SNG, 1980 tons of liquefied petroleum gas, 14 660 tons of light refinery liquid fuel, 8850 tons of solvent refined coal (the refinery's 'fuel'), and a further 2500 tons of this coal for sale. On a smaller scale, the COG refinery would manufacture 156 tons of valuable chemicals. The study showed that the net efficiency of the process would be about 75 per cent. The cost of the refinery might be around $800 million, with more than $600 million going on the refinery and $140 million on the coal mine to feed the refinery. As in most coal processing schemes, the complex would have to be built at the coal mine. If not, the project would miss half of the point, which is to turn coal into fuels that can be transported at costs far less than the cost of carrying coal. A breakdown of the cost of the COG refinery suggests that it compares favourably with the cost of developing new oil and gas fields to produce the same amounts of fuel.

The R & D effort devoted to coal processing and conversion is large and growing. Between 1961 and 1973 the Office of Coal Research spent $164 million on its programme. In 1975 federal expenditure on coal conversion R & D was $259 million, and a further $34 million from the OCR went on utilisation techniques such as fluidised combustion. In Britain the National Coal Board convinced a government study that the country should embark on a significantly larger utilisation R & D effort. Three projects that in 1974 were estimated to cost £20 million were given tentative approval. The projects were fluidised-bed combustion, coal liquefaction by solvent extraction, and pyrolysis. (The British Gas Corporation already had a sizeable coal gasification programme under way.) When the International Energy Agency made Britain the lead agency for coal R & D, the NCB revised its R & D package to make it even more ambitious. In November 1975 the IEA saw the first fruits of its attempts to promote international collaboration on R & D, when Britain, the US, and West Germany agreed to share the cost of building and operating an experimental fluidised bed. The $20 million project will be set up in Britain by a specially created NCB subsidiary – NCB (IEA Services) Ltd. This fluid bed will be a test rig rather than a prototype 'production' device. It will concentrate on the mechanical aspects of fluid beds, and will study the combustion of different types of coal as well as the retention of sulphur in the bed. The bed's exhaust gases will also be studied to see if they are compatible with gas turbine requirements.

The longer term

On the more distant horizon coal conversion technology could be linked with nuclear power so that the energy consumed in the various conversion processes comes from a nuclear reactor rather than from burning coal. The OECD report *Energy R & D* said of this idea: 'Such a technological breakthrough would reduce coal conversion costs by 20 to 30 per cent and increase the coal reserves available for conversion by a third.' Another 'two-fuel' system might be a coal-fired oil refinery. In both cases the aim would be to make better use of a scarce resource by enlisting the aid of a more abundant fuel. Another worthwhile venture might be a project to turn underground coal deposits into gas *in situ*, without digging up the coal. Not only would this eliminate the need to mine the coal before conversion, it would also allow the coal industry to exploit some reserves that just cannot be mined. This is another idea that has to be properly assessed before anyone acts and starts to spend money. These more speculative notions have not been reviewed as thoroughly as the more conventional coal R & D possibilities. With the coal industry facing a sudden change of fortune, it has not been able to look into every R & D option. Thus there is plenty to be done as far as project analysis is concerned, despite the veritable orgy of paper studies over the past couple of years.

Chapter 5

Oil resources

The doomsters may be wrong in their predictions that oil and gas (the fluid fuels) will begin to run out early in the next century, but this belief is so widespread that few governments or companies show much interest in spending massive amounts of money on large R & D programmes in this area. Clearly research can be relevant to the oil and gas industries; but no one would suggest that oil and gas R & D should take a share of the funds that matches their importance in the energy system as it is today. Thus the US's new Energy Research and Development Administration is spending more than ten times as much on coal R & D as it is on petroleum and natural gas. And in Britain the £4 million or so that the Department of Energy spent in 1974 on offshore oil and gas R & D was mostly in support of the government's statutory and regulatory functions, its environmental responsibilities, and the need for new technology to develop the North Sea oil and gas reserves.

Dixy Lee Ray's report WASH-1281 called for a significant programme of resource assessment. The OECD energy R & D report placed a similar emphasis, which it advocated for all energy resources: 'Energy R & D should start with a more systematic and continuous search for primary resources.' New exploration techniques are an important part of energy R & D and, 'if possible the ultimate goal of exploration R & D should be the discovery of methods for the direct detection of sub-surface oil or gas. Traditional methods, which can only detect the presence of geological structures favourable to oil or gas accumulation, are far from this goal. Perhaps "Bright Spot" techniques will be among the first direct detection methods.' Here the OECD report is referring to new techniques for interpreting seismic data. These may make it possible to find out from the seismic data whether or not those promising geological structures hold oil or gas, or are empty. At the present the only way to confirm the presence or

► The oil industry's move offshore in search of new oil and gas reserves has given birth to new and expensive technologies. British Petroleum's Forties field will produce 400000 barrels of oil a day (20 million tons/year). Four huge platforms will bring this oil to the surface and send it ashore by pipeline. Each platform cost around £100 million to build.

absence of hydrocarbons is to drill into the structure. This is a time-consuming and expensive business. And each unsuccessful hole only adds to the cost of any oil or gas that is found from successful drillings. (More often than not exploratory drillings prove disappointing.)

Once oil has been discovered there are some technological options that could improve the energy supply situation. For example, the yield of oil from a field can be boosted by new recovery techniques. And offshore oil and gas production, which is now only in its infancy, will come to dominate the oil and gas business as promising offshore prospects are developed. This move to less hospitable environments will continue as the rising prices of oil and gas make it profitable to invest ever larger sums of money in oil production.

Offshore exploration and production could, according to some oil men, yield more oil and gas than has so far been found on shore. The key to this massive wealth is the ability to work in deeper and deeper waters. As yet the industry has not gone much beyond 200 metres when producing oil. It is technically possible to drill at depths four times this, but production is a different thing altogether. As the industry moves into deeper waters it may have to abandon the platforms that now act as the interface between oil well and the pipeline ashore. The production equipment may have to be housed on the sea bed. The equipment needed for this is being developed, and it could be brought into widespread operation in the 1980s. According to the OECD report, there should be little difficulty in moving from 200 metres to 400 metres for oil production and 'full-scale offshore production through new systems might start in 1977, if present development is not accelerated. Later on, the technical possibility for descending to a depth of 3000 m exists, and the necessary hardware could be developed with an additional R & D effort.' The same report also points to the need for more knowledge of the marine and submarine environments: 'On the whole, offshore oil production will profit from, and in some cases will depend upon, further advances in all oceanological sciences. Certainly more R & D will be required to study climatic conditions, especially ocean currents, oceanfloor geology, marine biology, corrosion of different materials in the ocean environment and many related subjects.'

It isn't really necessary to find more oil to increase the world's oil reserves: perhaps the best way to improve the world's oil supply would be to squeeze more out of the existing fields. As it is, around two-thirds of the oil in a field is not recovered by existing production technology. (For gas the recovery is more like 80 per cent.) Recovery of the remaining two-thirds of the oil in a field hinges on the economics of oil production. New 'tertiary recovery' techniques are being developed, but they are more expensive than the straightforward business of drilling an oil well and letting internal pressure push the oil up to the surface. Natural pressure yields on average 20 per cent of the oil in a field. Secondary recovery pushes out the next 10 per cent or so. As in other activities in the energy research business, new recovery techniques have been boosted by the rise in crude oil prices.

Secondary recovery is a relatively straightforward process. Air, gas, or water is pumped into a well to force out more oil. This technique is already in wide use, often to maintain the flow of oil from a well as it begins to fall off. Secondary recovery has become much more popular since oil prices went up. Even this leaves large amounts of oil in a field. The National Academy of Engineering spelt out the magnitude of the 'wasted' oil in its report *US energy prospects: an engineering viewpoint*. This said, 'US oil reserves discovered to date originally contained approximately 430 billion barrels of oil. Primary recovery operations and conventional water or gas injection secondary recovery operations have already recovered or will recover about 140 billion barrels from these reservoirs, leaving some 290 billion barrels that cannot be produced by conventional methods.'

Water is sometimes pumped into a well as a secondary recovery technique, but it does not mix with oil; as a result a lot of oil remains clinging to the rocks in a reservoir. Chemicals could loosen the oil from the reservoir rock, allowing the oil to flow. Carbon dioxide and liquid petroleum gases are just two candidates for tertiary recovery additives. Steam is another possible substitute for water. (It will heat and thin the oil in a reservoir.) Clearly any chemical other than water will add to the cost of the oil obtained from a well. No matter how attractive it might be to squeeze out another 10 per cent of the oil in a field, this will happen only if the oil can be sold at a profit. The oil industry is re-evaluating what it can afford to spend to extract oil out of the ground. Tertiary recovery is beginning to look so attractive that the industry plans to invest more than $100 million to test new recovery techniques. One such test, in Arkansas, has been funded by the US Bureau of Mines which awarded $8 million to the Cities Services Company.

The techniques of tertiary recovery are such that different fields need different methods. Thus there is a lot of work to be done before tertiary recovery can be applied to every oil field. Some fields may, because of their rock formations, prove to be beyond help. In fact, some oil fields cannot be exploited with existing technology. These are the subject of other new recovery techniques. Extreme ideas, such as letting off a nuclear explosive in a well to break up the reservoir and release the oil, are being evaluated. These techniques are further off commercial exploitation than are tertiary recovery techniques. Indeed, such 'sledgehammer' approaches may be ruled out if they arouse significant public opposition, as seems very likely.

The NAE report puts into perspective the flow of oil that could be brought about by tertiary recovery: 'If a program of government assisted field tests is undertaken to overcome obstacles and implement enhanced oil recovery applications, it appears possible that these applications may yield 0.5 million barrels/day by 1980 and 0.8 million barrels/day in 1985 over what can be expected without the program.' (Britain expects to bring ashore 2 million barrels of oil a day from its North Sea fields in the 1980s.) The OECD energy R & D report warns that, while it may be possible to boost the recovery of oil from a well to 60 per cent, from its current 30 per cent, we will never get all of the oil out of an oil field. Even 40 per cent recovery would be a significant achievement as it would increase the recoverable reserves of oil by a third. A report from the United Nations Economic Commission for Europe pointed out that: 'The increase in production costs necessary to move from a, say, 45 per cent recovery ratio to 55 per cent, is so substantial that the opening up of new fields, in particular offshore, and the use of oil shale and tar sands seems more promising.' The report says that 'the average recovery factor for onshore operations lies at around 35 per cent for the ECE region as a whole. Off-shore operations have a slightly higher (40 per cent) efficiency.' According to the report, 'it would seem safe to expect that by the early 1990s the overall level of efficiency might not be beyond 45 per cent'.

Oil shales

Tar sands and oil shales have been all but untouched by the oil industry, and yet their reserves may be more than the reserves of conventional crude oil. The World Energy Conference survey of energy resources put recoverable resources of shale oil at 230 thousand million tonnes, while petroleum and natural gas reserves add up to 129 thousand million tonnes. As the WEC says, 'an accurate appraisal of the world's total resources of oil shales and bituminous sands is not possible' with existing data. Few countries have assessed their reserves of these unconventional oil resources. A survey of the shale situation in the UK said that there is a 'dearth of circumstantial information' on the extent of the largest shale deposits in the UK. The same report said that 'a case exists in the long term national interest for the precise extent and quality of the Jurassic shales in England to be assessed by a programme of geological investigation, including boring followed by some laboratory work'. The OECD energy R&D report says that 'systematic prospecting for oil shale has barely started but it is possible that the geographical distribution of oil shale is more equal than that of proven oil reserves. Searching more extensively for shale, especially for deeper than surface layers, should be a priority task.'

Oil shales are minerals containing an organic substance known as kerogen. This resinous hydrocarbon can be turned into an artificial crude oil by suitable processing. Exploitation of the world's shale resources requires the mining and processing of large quantities of shale, which can contain as much as 100 US gallons per ton of shale or as little as a few gallons per ton. In assessing the recoverable reserves of shale oil, it is generally assumed that shale containing less than 30 gallons of oil per ton will be uneconomical to exploit. (This clearly depends upon the cost of recovering the shale oil, and the prevailing price of crude oil.) A shale oil output of a million barrels a day might entail processing some 450–500 million tons of shale a year. In 1972 the coal mining industry in the US dug up 537 million tonnes of coal, split roughly equally between strip mining and deep mining. A shale industry would have to shift its material twice. When the oil has been extracted from the shale, the processor is left with a large pile of spent material. And he may find that the now useless material takes up more space than the freshly dug, oil-rich shale (shale can swell by a third during processing). Thus while shale deposits in the US may hold five times as much oil as the country's conventional crude reserves – perhaps 1.8×10^{12} barrels in the Colorado, Utah, Wyoming deposits – an enormous industry would have to be developed to exploit this resource.

Shale has to be processed to extract oil from it. This entails heating the mineral rock to 480 °C. At this temperature the kerogen is vaporised, and can be condensed separately as oil. Various processing techniques have been tried over the years. Scotland's shale oil industry, which dates back to the early 1860s, originally used manually filled and emptied retorts that were externally heated by burning coal. In the 1880s this horizontal retort gave way to a more sophisticated vertical retort. This gave semi-continuous operation. It also employed steam injection, producing ammonia as the hydrogen in the steam combined with nitrogen in the shale. The most advanced retort employed in the Scottish shale industry had a much higher thermal efficiency than its predecessors. This was achieved by injecting air as well as steam into the retort. As a result the carbon residue, left on the shale after the kerogen had vaporised, burnt, providing some of the heat for the process. Each retort was capable of a daily throughput of 12 tons. Britain's shale industry ceased operations in 1963, after more than 140 million tons of shale had been mined. British Petroleum was the last company to operate a shale mine in Scotland. The report on the UK shale outlook prepared for the Department of Energy concluded that there was no economic case for reviving the shale industry, but that it would probably have been operating profitably now had it not closed down in 1963. But its contribution to the country's oil supplies would be small: production would probably have been between 50 000 and 100 000 tons of oil a year. The situation in Britain is completely different from that in the US. While the UK's reserves of crude oil far exceed the shale reserves, the opposite situation appears to apply in the US.

▶ Shale oil is not a 'new' source of energy. Scotland's shale industry dug up 3 million tons of shale in 1910 and continued operating until 1963, thanks to tax concessions which allowed shale oil to compete with inexpensive crude from the Middle East.

A government report published in 1975 concluded that: 'If the Scottish shale industry had continued in the form in which it was operating in 1962, it would probably have been operating profitably at the present day by reason of the increased value of its products, but production would probably have been small', maybe somewhere between 50000 and 100000 tons of oil a year. The report found no economic case for re-opening the industry. There are, nevertheless, significant reserves of shale in Britain.

The photograph shows British Petroleum's Westwood shale processing works – it was taken from the top of one of the heaps of shale (known as 'bings') left over after the oil has been extracted.

Various shale processing techniques have been tried out by would-be shale oil companies in the US. In the 1950s the Union Oil Company operated a plant that processed 1000 tons of shale a day. The 'shale oven' was arranged so that an upward flow of hot shale met a downward flow of gas fron a gasifier. The shale oil condenses on the cool shale as it enters the bottom of the retort, and then flows through the oil outlet for subsequent processing. Union Oil's experiment was not commercially viable, but oil prices are now such that the company has embarked upon a second series of experiments and is designing a plant that could produce 50000 barrels of shale oil a day. If the company's plans are carried through, then the new unit could be operating by 1978. This plant would also upgrade the shale oil by a catalytic hydrogenation process.

TOSCO (The Oil Shale Corporation) is developing another technique for shale processing. Working on the principle that the heat content of a gas is too low to be able to heat the shale quickly enough – if the shale is 'cooked' for too long it tends to carbonise or gasify the kerogen – the TOSCO process uses a solid heat-exchange medium. Hot ceramic balls are mixed with preheated shale in a retort. The kerogen is vaporised, and the pebbles subsequently separated from the shale 'ash' for recycling.

There are other shale processing techniques, such as the 'Paraho' design of vertical retort. This is being developed by a consortium of oil companies. The $7.5 million Paraho project will see if the prototype kiln design, which can handle 470 tons of shale a day to produce 300 barrels of oil, can be scaled up to a commercially viable size. Ultimately 10 to 12 such units might be used in a commercial operation producing 100000 barrels of oil a day. In the Paraho kiln shale flows downwards, and is heated to 600 °C as it does so. The kerogen breaks down into a vapour; when this is cooled shale oil and gas are separated. Paraho's plant would also upgrade the oil from the shale, thinning the oil and removing nitrogen from it.

Any 'dig and retort' technique of shale exploitation inevitably involves massive mining operations. It may be possible to establish a profitable business, but there are significant environmental drawbacks. Not only does shale recovery, which will be by open-cast mining to begin with, disturb large areas, it also produces immense quantities of waste material. A shale operation producing a million barrels of oil a day would dig up between $1\frac{1}{4}$ and $1\frac{1}{2}$ million tons of shale a day. More than a million tons of spent shale would be left over after the kerogen had been extracted. And this spent shale would take up more space than the raw material, so it could not be disposed of simply by refilling the original diggings. Thus an industry producing just 5 per cent of American oil needs would have to dig up nearly half a billion tons of shale each year, process it to extract the oil, and then find an acceptable way of disposing of the spent shale.

One way round the enormous environmental problems might be to extract the oil without moving most of the shale. This would be possible if the oil could be extracted from the oil shale where it is found. This is known as in-situ recovery. While one motivation for developing in-situ recovery is environmental, economic factors also provide a stimulus. With conventional recovery techniques the processes that take place between getting the shale out of the ground and pouring the oil into a barrel account for more than half the cost of shale oil recovery.

One in-situ recovery technique involves drilling a number of wells into a shale formation. If the hydrocarbons in some of the holes are ignited and air is pumped down to the oil, a combustion front could spread through the shale formation. The spreading flame would heat the shale and release some of the oil, which could then be recovered through other wells. Before a flame could spread through the shale bed, the rock would have to be broken up. An extreme approach to this problem would be to let off a nuclear explosive in the shale. Other ways of breaking up the shale to make it permeable include the technique proposed by Occidental Petroleum. To begin with, a hole is dug at the bottom of a shale formation – this requires some conventional mining. Occidental then suggests that the shale above the hole should be broken up by explosives so that the fractured rock would collapse into the mined space. If a hole is drilled down through the fractured shale, air or gas can be pumped down into the bed. If combustion is set under way by injecting natural gas into the well, the shale will be heated and will flow out of the rock. As combustion proceeds it leaves behind a deposit of carbon and eventually the flow of natural gas into the well can be replaced by a flow of air, allowing the carbon to burn thus generating enough heat to maintain the process. Oil produced during the combustion flows down to the bottom of the well: it can then be pumped to the surface.

In-situ mining of shale looks like a good way out of the environmental difficulties, but it too could have undesirable environmental impacts. Not enough work has been done to show whether or not it will lead to subsidence of the land above the operation. The technique also might contaminate underground watercourses. These questions are far from answered, and it will be some time before in-situ shale exploitation becomes viable. Indeed, it may prove to be impossible to extract oil from shale without digging up the material.

In its report on oil shale, published in 1973, the US National Petroleum Council concluded that 54 billion barrels of synthetic crude might be economically recoverable from the Colorado, Utah, Wyoming deposits – the Piceance basin. The NPC's analysis showed that, depending on the rate of return required from the original capital investment, synthetic crude might cost between $3.90 and $6.35 a barrel. Clearly the spate of inflation that has hit the world since 1973 will have pushed up the price, and the prevailing high rates of interest mean that shale oil costs will be towards the top of the NPC bracket, but oil shale now seems to be in a reasonably attractive position. A more recent assessment of the costs of crude oil from different sources in the mid-1980s was presented by the OECD in its report on energy prospects to 1985 (prices are in 1972 dollars per barrel):

Persian Gulf	0.15–0.20
US:	
Low cost category	0.30–2.60
Medium cost category	3.30–6.70
High cost category	above 7.00
North Sea	1.50–2.00
High grade oil shales	4.11–7.30
Tar sands (Canada)	3.40–3.80
Syncrude from coal	6.50–7.50

Apart from showing that the *price* of oil in world markets has got little to do with the *cost* of the oil, this table does show that shale oil can be cheaper to recover than some crude oil. And it should be possible to recover shale oil at costs below the market price of oil, as fixed by the oil-producing countries.

The economics of shale exploitation must have been reasonably attractive at the beginning of 1974 when, in January of that year, the Standard Oil Company (Indiana) and the Gulf Oil Corporation put in a winning bid of $210.3 million for 5089 acres of government-owned shale land in western Colorado. (The US government owned 8.3 of the 11 million acres of shale territory before it began auctioning it off.) Standard and Gulf were expected to spend $500 million to build a 100000 barrel per day synthetic crude plant. This cost figure was originally quoted in 1971 by the National Petroleum Council and was confirmed in 1974 when a technical progress report from the US Bureau of Mines put the cost of a 50000 barrel per day plant at $279450100 and a 100000 barrel per day plant at $522375400. On top of these prices you have to add the cost of land.

The Standard/Gulf bid was followed a month later by an $118 million bid by Atlantic Richfield for 5100 acres. The Standard/Gulf plot was estimated to contain between 1300 million and 4000 million barrels of oil, depending upon whether or not underground mining or surface mining was used to extract the shale. The Atlantic Richfield land was estimated to hold 700 million barrels of oil. A third bid, of $76 million from Phillips Petroleum and Sun Oil, secured another 5100 acre plot, with estimated reserves of 244 million barrels. While hardly negligible, the price, implicit in the bids, attached to each barrel of oil was small in comparison with the estimated 'break-even' price for economic shale recovery operations.

After the first flurry of excitement had died down the oil companies began to calculate just what was involved in developing the then non-existent shale oil industry. The costs for shale operations look high when set alongside the cost of developing offshore oil reserves. At the beginning of 1974 it was reckoned to cost £1500 (about $3500) to bring into operation each barrel per day of production capacity; during the year this rose to £2000, with the anticipation that new fields would not be brought on line for less than £3000 per barrel per day of production capacity. Leaving out the cost of land, which is far from insignificant, $500 million for a 100000 barrel per day shale oil plant works out at $5000 per barrel per day. And this is for a technology that has not been proved and whose ultimate development costs can only go up (to way above the original estimates if previous experience of technical development is anything to go by). On top of this, fears about environmental damage, and the likely shortage of the water needed to develop the industry, made the outlook gloomy almost before the original ripples of euphoria had died down. (A report by the US Geological Survey concluded that there was only enough water available in the Green River shale formation to support the first two 5100 acre plots, and that to develop the rest of the huge deposits would inevitably require massive imports of water into the area.)

The prospects for a thriving shale industry were not brightened when, in October 1974, Colony Development Operations suspended work on its Colorado shale processing plant. Construction of the facility was to have started in 1975, but in 1974 the company found that the original estimate for the cost of setting up the 50000 barrel per day operation had risen from $450 million to $800 million. The company (a joint venture between TOSCO, Atlantic Richfield, Ashland Oil, and Shell Oil) decided that the outlook was too doubtful to justify continuation of the project.

Another doubt was raised about the viability of shale oil recovery. Some experts maintained that a shale oil processing plant might consume more energy than it produced. That is, each barrel of shale oil from the plant would, so they said, be produced only if more than the equivalent of a barrel of oil went into the operation. This 'energy analysis' of shale oil recovery would, if true, make it very difficult to justify the whole business. It would only make sense if the energy input were in a less valuable form than the output. Coal, for example, might be used to heat the shale oil recovery equipment; but as we have seen there are other ways of turning coal into oil. The man who invented the Paraho process, John Jones, who is also the president of the Paraho Development Corporation, denies that Paraho's shale operation would be a net consumer of energy. According to him, Paraho can recover enough gas from the shale processing plant to heat the retorts, with some left over to generate electricity. It may be possible to gain energy from a shale operation, but this sort of argument clearly shows that energy projects are now subjected to a thorough evaluation before they are accepted as viable.

Tar sands

There are various estimates of the oil reserves tied up in the Athabasca tar sands of Canada's Alberta province. One estimate has put the world's tar sands reserves at 225 billion cubic metres, which is equivalent to 1600 billion barrels of oil. The World Energy Conference Survey listed the recoverable reserves as 130304 megatonnes (955 billion barrels, assuming 50 per cent oil recovery). A Canadian report on Alberta's tar sands reserves said that 'The Alberta Energy Resources Conservation Board considers that proved recoverable (mineable) reserves of crude bitumen currently stand at 38 billion barrels of synthetic crude oil...The Board has further estimated that the ultimate amount of crude bitumen in place within the presently delineated deposits is roughly 1000 billion barrels, using a cut-off grade of 2 or 3 weight per cent crude bitumen. Of this total, 330 billion barrels of bitumen is considered ultimately recoverable, corresponding to a volume of synthetic crude oil of approximately 250 billion barrels.' As with all figures for energy reserves, the tar sands estimates have to be viewed with caution. They can do little more than establish a reasonable 'order of magnitude' estimate. We can see from these numbers that the tar sands deposits are hardly chicken-feed when set beside the WEC's 1974 estimate of 671 billion barrels for the world's crude oil recoverable reserves.

As with shale oil, economical recovery and extraction of the oil from tar sands is the major technical problem to overcome before the industry really takes off. The OECD estimates of the cost of crude oil from different sources show that the oil from tar sands should be significantly cheaper than oil from shale (see p. 83). In fact, a sizeable tar sands plant is already operating in Canada. Great Canadian Oil Sands (GCOS) has been operating a synthetic crude plant on the Athabasca tar sands since 1967. This plant can produce 45000 barrels of oil a day, and there are plans to expand the plant. As with any tar sands plant, GCOS has to upgrade by hydrogenation the bitumen that is extracted from the sands. This produces a synthetic crude that can be fed into a refinery. GCOS spent $500 million and accumulated losses of $100 million or more before the increase in oil prices put the company into a profitable situation.

As with shale oil, it takes a massive effort to extract synthetic crude from tar sands. Here too exploitation of a massive source of energy depends upon setting up a huge mining industry. However, it is easier to extract oil from sands than from shale. Tar sands contain an average 12 per cent bitumen. The sands are easily enough treated to remove the bitumen. The viscosity of the bitumen falls rapidly as it is warmed, hence it can be extracted by heating the sands and washing the oil out with hot water. GCOS employs this process.

The problems of operating a large mining industry in the extreme conditions of the Alberta winter make tar sands extraction a formidable task. With temperatures falling extremely low, the sands become a viciously abrasive substance that can wear away steel digging equipment in next to no time. This and the impossibility of exploiting much of the tar sands reserves with conventional mining techniques have catalysed the pursuit of in-situ recovery techniques. Shell Canada Ltd and Shell Explorer Ltd, a subsidiary of the Shell Oil Company, are developing a steam-based process which, according to Shell, 'involves a conventional steam drive followed by pressurisation and depletion cycles'. A cycle of steam injection and pressurisation should make the oil in the tar sands flow as it does in a conventional reservoir. In this way efficient oil extraction can be achieved without having to dig up the sands. Shell estimates that its pilot tests, which were due to start by mid-1976, could cost $33 million. In-situ recovery may make it possible to exploit more of the tar sands deposits, but only a third or half of the oil will be extracted. This is much less than is obtained by the mining techniques now in operation. Other possible extraction methods include solvent recovery. Dr Alfred Globus, president of the Guardian Chemical Corporation, has claimed that 95 per cent of the bitumen could be got out of the sands by employing a solvent dissolved in water. Another alternative is underground combustion followed by recovery of the oil through wells. All of these possibilities are very tentative and require evaluation before they can be adopted or ruled out. Early in 1974 the Province of Alberta set aside $100 million for a five-year programme of tar sands R&D.

Tar sands may have an edge over oil shale when it comes to the cost of recovering oil, but the development of a sizeable tar sands

▶ The Athabasca tar sands may be a rich oil deposit, but the industry has to operate in a far from friendly environment. In winter the temperature can be so low that metal machinery becomes brittle, forcing mining operations to stop.

operation is anything but a foregone conclusion. The most ad-vanced new project is the Syncrude venture, which is run by a consortium made up of various oil companies. The membership of this consortium has changed during the life of the project and at one time the whole venture was in doubt when several members of the consortium pulled out. Only the intervention of the Canadian federal government and local state governments stopped the whole project from collapsing. Syncrude decided to go ahead with its plans for a plant that could produce 125 000 barrels of oil a day after 15 years of pilot operation costing $55 million. The Syncrude operation will involve shifting 92 million tons of tar sands each year, with a daily load of 275 000 tons passing through the separation plant. This venture was to cost $1 billion when the project was first mooted; in December 1974 ARCO pulled out of the project because the cost had risen to $2 billion.

The motivation for Canada's development of the tar sands is clear. In a report on Alberta's tar sands Dean Clay says that the oil sands are 'a key element in determining Canada's intermediate-term oil supply'. He adds that 'potentially serious problems of domestic oil supply [will] arise around 1980 or later, and... Canada is liable to lose its net self-sufficiency in oil early in the next decade'. The rising cost of the Syncrude project must slow down the expansion of the tar sands industry, but Clay's report concludes that 'if no in-situ development occurs by 1985, the probable maximum production from the anticipated four to five mining ventures would be 500,000 to 600,000 barrels/day'.

The prospects for the unconventional oil industry are reason-able if not spectacular. Tertiary recovery of oil from underground reservoirs begins to look more viable as oil prices rise; offshore exploration and production is going from strength to strength; oil shales and tar sands may have their ups and downs but both are likely to develop in a healthy economic climate. (The synthetic crude business has had some difficulty in raising funds for large facilities.) And yet all of these technologies are far from commer-cial reality. This state of affairs owes much to the organisation of the oil business. Oil is not supported by an orderly and widely known R&D effort. With a few exceptions there is a minimal public effort in the oil sector. Of the European countries France stands out, with the Institut Français du Petrole – a state-run research institute funded by a tax on the country's oil consumption. In Britain the government has tried to strengthen its oil R&D activity by strengthening the Marine Technology Support Unit (MATSU), at Harwell, and by establishing an Offshore Energy Technology Board (OETB). But most of the work is done by the oil companies, and they are notorious for their secrecy – they don't like telling anyone what they are doing, let alone what the results of their work are. Now that governments have been thrown into closer contact with the oil companies, we can expect to see some changes in the freedom of the oil companies to pursue what is in their best interest rather than the interest of the countries in which they operate.

Chapter 6

Nuclear fission

Nuclear power, energy's great white hope, is surrounded by some of the greatest doubts and difficulties of all new energy technologies. The people in the nuclear business might not agree with this assessment of their problems, but no other energy system is the subject of a sustained and sometimes bitter campaign to bring it to a halt. Over the next few decades nuclear power could take over the generation of electricity and relieve fossil fuels of this burden. But nuclear power has been severely criticised on safety and environmental grounds; and there are claims that the energy needed to establish a nuclear power programme is so high that it worsens the energy supply situation before it eases it.

The nuclear safety issue is complex: it is not solely technical. A catalyst for the growth of the anti-nuclear movement in the US was the way in which the US Atomic Energy Commission (AEC) behaved during the 1950s and 1960s. As both an active promoter of nuclear power, and the body set up to protect the public from nuclear hazards, the US AEC was trying to fulfil two mutually exclusive roles. There is enough evidence to show that the US AEC was a less than thorough watchdog; and that it behaved with excessive secrecy. Something has been done about this unsatisfactory situation: the US AEC has been dismantled and its two roles have been vested in two new organisations. Nuclear research and development is just one subject, albeit the major one, for the newly formed Energy Research and Development Administration to handle. Nuclear safety and related environmental issues are now the responsibility of the Nuclear Regulatory Commission.

Britain once had a similar conflict in its nuclear affairs. The man whose job it was to advise the government on issues of nuclear technology was also the head of the Nuclear Installations Inspectorate. A sequence of events led to a change in the organisation of

Britain's nuclear regulation. To begin with the House of Commons Select Committee on Science and Technology published a report in which the undesirability of this concentration of conflicting powers was brought into the political arena. And when the government decided to set up a separate Department of Energy, the new energy minister took the opportunity to appoint a new chief scientist. And not long after the new department was set up all safety and environmental responsibilities, nuclear or otherwise, were brought together in a new organisation.

No matter how much governments reorganise their nuclear bureaucracies, they will never erase the stigma of the 'bomb'. Nuclear weapons are a massive millstone around the neck of the peaceful atom. The first reactors were built as a part of national weapons programmes, and Britain owes its early lead in nuclear technology to its desire for a nuclear capability. And there is no denying that the fissile materials that fuel a power station reactor *can*, with the right treatment, be turned into weapons material.

The weapons issue is something of a red-herring when set alongside the nuclear safety issue. In its report on energy prospects to 1985 the OECD put the case for greater public information: 'It is obvious that if the public believes that there are gaps or weak points in nuclear industrial safety, the execution of nuclear programmes might be delayed. This confirms the importance of promoting wide information on nuclear activities, and the priority which should be given to safety requirements in preparing decisions on energy policies.' Some countries are learning this lesson faster than others. The US, with its constitutional emphasis on the freedom of information, has responded more quickly than most European countries, including Britain where there has never been a tradition of public access to information.

Thermal reactors

A look at the reserves of fissile and fertile nuclear materials shows that nuclear power can, with the right technology, see us through the next century or so. Fissile material, such as uranium-235 or plutonium, can be 'burned' in a variety of reactor configurations. A reactor is conceptually simple enough. All you have to do is assemble the fuel in the right amount and the right configuration, intersperse it with a moderator that slows down the neutrons and makes them more amenable to fission reactions, and the 'pile' will get hot. This heat, which can be removed by a flow of coolant, is what nuclear power is all about. Ultimately it is turned into steam to drive a conventional steam turbine electricity generator.

Fertile material such as uranium-238 is not useful as a reactor fuel, but nuclear reactions that take place inside a reactor can convert it into fissile material. For example, uranium-238 can be converted into plutonium. This conversion takes place in all reactors but it is more rapid in a fast reactor; hence these are also called 'breeder' reactors because they can breed fissile material from fertile material.

The technical problems that go with turning this basically simple concept into a power station are numerous and complex. To begin with there are various permutations and combinations of coolant and moderator – the two can be the same thing, as in the Canadian CANDU reactor which has heavy water as both coolant and moderator, and which also has the advantage that the fuel does not have to be 'enriched'. (We will see later in this chapter that enrichment is just one of the parts of the nuclear fuel cycle that could hold up the development of nuclear power.)

There is little evidence that the reactors now on sale by the world's nuclear manufacturers were designed for their energy efficiency or their technical superiority, let alone their greater safety and environmental excellence. Today's reactor types did not exactly come about accidentally, but they are the result of a series of choices and chances. America's light water reactors (LWRs), for example, stemmed from submarine propulsion units. That this type of reactor, in its several manifestations, is the most numerous is a result of the sheer size of the American commitment to the LWR. (At the end of 1974 the US had 55 plants with operating licences, 62 with construction permits, 107 reactors on order, and 12 letters of intent, giving a total capacity of 233 817 MW (electrical).) Presented with the availability of the nearest thing you can get to an 'off-the-shelf' reactor, other countries have adopted the LWR. Following the oil price rises, in 1974 the French government embarked upon a huge nuclear programme, based on LWRs, that could meet more than 90 per cent of the country's electricity requirements by the turn of the century (a capacity of something near 170000 MWe).

At one time it looked as if Britain might adopt a similar line, but in the end the long and drawn out arguments about the UK's nuclear future resulted in yet another type of reactor being chosen for the next stage of Britain's nuclear programme. The Steam Generating Heavy Water Reactor (SGHWR) is cooled by ordinary water and moderated by heavy water (deuterium oxide).

The different types of reactor vary little in their nuclear reaction. All rely on fission of uranium-235 nuclei (and possibly plutonium nuclei) sparked off by neutron collisions in a sustained chain reaction. To make these collisions more likely, the reactor's moderator slows down (thermalises) the fast neutrons produced by fission reactions.

The uranium dug up in the world's uranium mines holds just 0.7 per cent uranium-235. For the next two or three decades uranium-235 will be the essential nuclear fuel; it will, therefore, be the limiting factor in the nuclear fuel cycle. In 1973 the Nuclear Energy Agency (NEA) estimated that there might be more than 1000 GW (electrical) of nuclear capacity installed by 1990 – more than ten times the current world figure. Since then, of course, many countries have upped their nuclear plans. The OECD's report on energy prospects to 1985 estimated that an accelerated programme of nuclear power might take the OECD nuclear capacity in 1985 to 700 GW (electrical) from an earlier estimate of 513 GWe.

In 1973 the NEA estimated that 'annual demand for uranium is expected to establish itself in the region of 60000 tonnes uranium by 1980 and almost double this figure by 1985'. The rapid growth of nuclear power could cause problems:

'No shortages of uranium supply are to be expected in the 1970s. However, the rapid growth in demand in the coming decade cannot be satisfied on the basis of existing uranium exploration levels. Given the necessity of a lead time of about eight years between discovery and actual production, it is therefore essential that steps be taken to increase the rate of exploration for uranium so that an adequate forward reserve may be maintained.'

Another way of putting this is that any power station ordered now will be fuelled by uranium that has yet to be discovered. This is

not as speculative as it seems, but it does bring home the importance of uranium exploration activities.

The WEC survey of energy resources says that 'total world
resources in the non-Communist nations recoverable up to $39 per
kilogramme of uranium are estimated to be approximately 4.0
megatonnes. Nearly half the resources are reasonably assured
resources or reserves.' So far, uranium mining has not worked
reserves at this price; but prices are rising. 'Rapidly rising prices
were the dominant element of the uranium market in 1974, as
prices for delivery in the years 1975 through 1980 rose an eye-
popping average of 105 per cent. By year's end purchasers were
bidding $15 per pound U_3O_8 for immediate delivery and $25 for
delivery in 1980.' (These prices are equivalent to $39 and $65 per
kilogramme of uranium.) This summary of the uranium market,
made by Jack Mommsen of the Nuclear Exchange Corporation,
shows that market forces are already beginning to affect the
situation for the good. The rush into nuclear power puts pressure
on uranium prices, making it more attractive for people to seek
out and exploit new ore regions.

There is one massive source of uranium that occasionally crops
up when this subject is discussed. According to a study carried
out by Britain's Central Electricity Generating Board (CEGB),
'the oceans contain 4×10^9 tonnes of uranium'. This is dispersed
pretty uniformly at a concentration of 3 parts per billion. The
CEGB study came to a number of conclusions:

'The extraction of uranium from seawater around the UK
could not meet the UK's requirement for uranium fuel for an
expanding thermal reactor programme.

'In order to make a significant contribution to the world supply
of uranium a seawater handling scheme employing pumped flow
must be considered.

'A process for the extraction of uranium from seawater has not
yet been fully developed.

'Uranium from seawater is likely to cost at least $70 per pound
at 1974 prices.'

Another fly in the ointment is the energy cost of extracting
uranium from seawater. The CEGB does not mention this in its
study, but other studies have shown that with thermal reactors,
more energy may have to be put into extracting the uranium from

Table 6.1. *Typical LWR fuel cycle costs (1972)*

	Cost in m$/kWh	% of total	Corresponding costs
Uranium	0.50	24	$8/lb U_3O_8
Enrichment	0.47	22	$32/kg separative work
Fuel fabrication	0.33	16	
Services (reprocessing, transport, etc.)	0.17	8	
Indirect costs	0.63	30	
Total	2.10	100	

seawater than is available from the uranium. It is generally true that less rich uranium ores will require higher energy inputs to mine and process them. It does not do to ignore this fact of life when uranium reserves and resources are discussed.

Another important factor that we must not overlook in this talk of uranium reserves and prices is the relative unimportance of uranium prices in the cost of electricity from nuclear power stations. Table 6.1 shows the fuel cycle costs for an LWR as presented in the OECD report on energy prospects to 1985.

This is perhaps a good time to look at some of the other economic factors that are significant in the nuclear fuel cycle. Table 6.2, from the same report as the table showing fuel cycle costs, shows the typical investment figures for the whole nuclear cycle. In the table all of the costs of a power station are lumped in together. A breakdown of this figure shows that the nuclear part of the power station – the nuclear steam supply system – costs less than the conventional, non-nuclear parts. In its report 'on the choice of thermal reactor systems' Britain's Nuclear Power Advisory Board wrote that 'the reactor section of a nuclear power station represents about 35 per cent to 50 per cent of the total construction cost of the station, depending on the system. The non-nuclear parts of a station cost much the same for all systems.' The NPAB published the following table (table 6.3), which was prepared by the CEGB, in its report. Any tables such as these have to carry a word of warning about the effects of inflation, especially in these days of economic instability. The important point is not the absolute

Table 6.2. *Typical investment figures in the fuel cycle industry to sustain a 1000 MWe LWR*

	Capacity	Specific investment	Total investment ($m)	% of total investment
LWR	1000 MWe	$300/kW	300	91.2
Uranium mining	120 tU/yr	$33/kgU/yr	5	1.5
Conversion to UF$_6$	120 tU/yr	$5.6/kgU/yr	0.8	0.24
Enrichment	90 tSWU/yr	$200/kgSWU/yr	20	6
Fuel fabrication	21 tU/yr	$50/kgU/yr	1.2	0.36
Reprocessing and waste management	21 tU/yr	$60/kgU/yr	1.6	0.5
Total (rounded)			329	100

tU = tonnes uranium; tSWU = tonnes separative work units; kgU = kilogrammes uranium; kgSWU = kilogrammes separative work units.

Table 6.3. *Estimates of capital and lifetime costs for nuclear reactors*

	Magnox	AGR	HTR	SGHWR	PWR
Construction costs (including nuclear steam supply system)	241	171	142	150	132
Nuclear steam supply system cost	(116)	(89)	(60)	(67)	(50)
Interest during construction	72	51	42	45	40
Initial fuel	15	19	13	18	14
Total operating costs	38	52	63	49	47

Costs per kilowatt sent out (£)

figures, but the difference between the various parts of the nuclear cycle. By now all of these numbers may have doubled, but the relativities are less volatile. The CEGB produced this table to show that the reactor it wanted to build, the pressurised water reactor (PWR), was cheaper than the alternatives – the SGHWR, and the various gas-cooled reactors, Magnox, the Advanced Gas Cooled Reactor (AGR), and the High Temperature Gas Cooled Reactor (HTR). However, at the time it was pointed out, particularly by the CEGB's critics, that the inaccuracies in these cost estimates were far bigger than the differences between reactors, making it very difficult to choose a reactor system on this basis alone.

Breeder reactors

The thermal reactor, in its various designs, is but one nuclear option. Uranium reserves will certainly turn out to be larger than they now are, but not so much so that thermal reactors alone could keep society 'electrified' for very long. Indeed, so short would be a purely thermal era that nuclear power would hardly be worth the effort that has been lavished on it. It is only the presence of the so-called 'breeder' reactor on the not too distant horizon that makes all the effort worthwhile. Whereas thermal reactors can extract something like 1 per cent of the energy that is tied up in the uranium dug out of the ground, the breeder might extract 60 per cent of the energy locked in the uranium nuclei. The breeder would achieve this gain by putting to use the uranium-238 which makes up 99.3 per cent of natural uranium, and can be converted into plutonium. The way to do this is to throw out the moderator that is put into thermal reactors to slow down the neutrons produced by fission reactions. If the neutrons remain 'fast' some of them take part in fission reactions and others are captured by uranium-238 nuclei to breed plutonium. Similar reactions occur in thermal reactors, but at nothing like the uranium conversion efficiency of a fast breeder reactor. The core of a breeder can be designed so that it is converting uranium-238 into plutonium faster than it is burning up uranium-235 and fissile plutonium fuels. If the 'doubling time' – the time it takes for the breeder to convert enough uranium-238 into plutonium to fuel another reactor – is short enough, the demand for freshly mined uranium-235 could be reduced drastically, thus stretching nuclear fuel reserves so that they could last far beyond the beginning of the 21st century.

The reasons for developing the breeder reactor are clear; but the path towards a breeder economy is not so straightforward.

Britain, France and the Soviet Union took an early lead in this field. The US's committee approach to its Clinch River liquid-metal fast breeder reactor (LMFBR) project has caused problems because of the American desire to involve industry as early as possible. This meant that the Clinch River project was carved up between different companies, with the Breeder Reactor Corporation managing the project. This committee approach to the programme should take some of the blame for the massive inflation that hit the project even before work had begun on the reactor site. High price and safety worries brought the breeder considerable criticism, which the Energy Research and Development Administration responded to by retrieving management control of the project.

Whatever becomes of the breeder project, it will be some time before fast reactors are a commercial proposition. The most popular approach to fast reactors is the LMFBR. This has the fast reactor's dense and hot core cooled by a flow of molten sodium. Liquid metal technology is not the easiest to develop, especially when there are heat exchangers in the system. These transfer the reactor's heat from liquid metal circuits to steam circuits. A leak in a hot exchanger could lead to an explosive reaction between steam and sodium. The first fast reactor of any size to be completed – the Soviet BN350 reactor – was damaged by water/metal explosions in leaking heat exchangers. Britain's Prototype Fast Reactor (PFR) also suffered from small heat exchanger leaks.

The current generation of fast reactors are all about the same size: both the French and British LMFBRs have a nominal capacity of 250 MW (electrical), the Soviet BN350 is a dual purpose desalination/electricity generation unit producing 150 MW (electrical), and the Clinch River LMFBR, which should start operating in 1982 or soon after, will produce 400 MW (electrical). These are by no means small reactors, but commercial fast reactors will probably have a generating capacity around 1300 MW (electrical). The Soviet Union is building a 600 MW (electrical) fast reactor (BN600); France is developing a 1200 MW (electrical) reactor (Super Phenix); Britain plans to build a similar sized reactor (the Commercial Fast Reactor, CFR). None of the larger units, with the exception of BN600, will be operating before 1980. Commercial fast reactors will not be ordered in any numbers before the early 1980s. By this timetable, the fast breeder reactor cannot make a

▶ Fast reactors are an essential part of many energy R & D programmes. The
Prototype Fast Reactor (PFR) started operating at Dounreay in Scotland
in 1975. The photograph shows the inside of the reactor, partly loaded
with a dummy core, neutron reflector, and neutron shield.

significant impact on the world's energy picture before 1990 at the earliest. Until then we shall have to live with thermal reactors.

There are some nagging doubts about the direction of the world's breeder reactor programmes. All are pursuing the same basic concept, the LMFBR. This is by no means the only design that can be developed; and there are some nuclear technologists who maintain that the LMFBR is the wrong breeder. Its doubling time, they say, is too long. It is true that the prototype fast reactors are not true breeders – they convert uranium-238 to plutonium, but not as quickly as they consume uranium-238. The LMFBR can be made into a positive breeder, but the fear is that the final commercial designs will have a doubling time of 25 years or more, which is hardly short enough to relieve the uranium supply situation.

The aim of any breeder reactor programme is to halt the growing need for uranium, and to multiply by a factor of 50 or so the energy that can be obtained from a quantity of uranium. This means developing a breeder fuel cycle with a doubling time that allows uranium-238 to be turned into plutonium rapidly enough to reduce the need for uranium-235 from newly mined uranium. The whole breeder fuel cycle determines the fissile material 'doubling time' – how quickly the irradiated breeder fuel can be reprocessed to recover plutonium, for example – but the key factor is the plutonium breeding rate in the reactor. Today's prototype fast reactors are hardly breeders – they were not built to make plutonium but to prove the reactor concept – and it seems that an LMFBR with the same fuel as these prototypes would have a doubling time of 25 years or more. To bring the doubling time down to a more respectable 15 years or so would require the development of a completely new type of fuel – based on uranium and plutonium carbide rather than the uranium and plutonium oxide of the prototypes.

An alternative approach to the LMFBR with carbide fuel is the gas-cooled breeder reactor (GCBR) with a more conventional oxide fuel. The GCBR has a potentially better breeding rate because the coolant – probably helium – would 'soak up' and waste fewer neutrons so that there would be more of them available for fuel breeding reactions in the core. According to R. D. Vaughan, a British nuclear engineer: 'The gas-cooled breeder reactor, using mixed-oxide fuel and with engineering development based upon 20 years worldwide experience with gas-cooled reactors, offers a good prospect of achieving the necessary breeding performance.'

The LMFBR advocates come back with mutterings about the safety of the GCBR, and the difficulty of ensuring that its coolant never escapes. With the GCBR's hot dense core it would certainly be a catastrophe if the coolant ever disappeared from the core; but concrete containment systems have a pretty reliable record and arouse few of the doubts that go with some containment systems. In any case, if the coolant ever escaped from a GCBR core, the nuclear reaction would cease; whereas if the LMFBR lost its coolant the reaction rate could increase. It may seem that the breeder reactor arouses enough controversy without arguing about rival reactor types, but it is important to get the breeder reactor strategy right. The alternative options should be evaluated thoroughly before we have passed the point of no return. Should we, for example, opt for the GCBR in tandem with the high temperature gas-cooled reactor (HTR)? This would give us a wholly gas-cooled system; but there are criticisms of the HTR because it produces more severe radioactive waste problems than other reactors. Does this mean that we should leave the HTR undeveloped and rely on a GCBR to provide us with hot gas for process heat?

Any radical change in approach would disturb the existing breeder reactor programmes, all of which rely heavily on the LMFBR concept. So it is difficult to see the nuclear establishment ditching the LMFBR in favour of another system. As it is, LMFBR development is proving to be such a costly business that the smaller countries working on this system are considering the possibility of international cooperation. The added cost of yet another system might be too much, so we can't be sure that the best option will be adopted.

The fuel cycle

The complexity of the nuclear cycle is such that there are several key steps outside the reactor. The major additional steps are uranium mining and mechanical processing, chemical processing of uranium, uranium enrichment, fuel manufacture, and the reprocessing of irradiated fuel that has done its turn in a reactor. All are sizeable operations requiring significant investment (see p. 97). Uranium enrichment has, in the past, been viewed as the key link in the fuel chain; however, nuclear power depends upon all of the links being present. In 1974 it became clear that fuel reprocessing might be a key element in the growth of the American nuclear power programme. In its 1974 report on the US nuclear industry the AEC said:

'In July 1974, the General Electric Company (GE) advised the AEC that the reprocessing plant it had been constructing at Morris, Illinois, definitely would not begin operation in 1974 as previously planned and that, unless the plant was redesigned and rebuilt, it could not be operated at all. GE estimated that the necessary modifications might cost between $90 million and $130 million and would require four years of work.'

GE's problem was that it attempted to incorporate a new and untried process in the plant. With nowhere to send their used fuel elements, the electricity utilities expressed fears that they would run out of space in which to store spent fuel.

If fuel reprocessing can cause problems at the tail end of the nuclear cycle, the key technological element at the beginning of the cycle is uranium enrichment. This is the process by which the uranium-235 concentration of uranium is boosted from its natural level of 0.7 per cent to a more useful level about three or four times higher. Because the two isotopes of uranium are chemically identical, separation is by a physical process. So far enrichment has been by gaseous diffusion. Uranium is turned into a gas (uranium hexafluoride, UF_6) and diffused through a series of porous barriers – the lighter uranium-235 diffuses somewhat more rapidly and the gas that seeps through the barriers is slightly enriched in uranium-235, leaving behind a gas depleted in this isotope. Uranium hexafluoride, which is not an easy gas to handle, has to be passed through about 2000 diffusion stages to achieve the required degree of enrichment. The gaseous diffusion process is very energy intensive, with America's diffusion plants a significant drain on the country's electricity supply (around 6000 MWe to run three diffusion plants).

New enrichment technologies are being developed in an attempt to cut the size and cost of enrichment plants, and to reduce their energy consumption. The European gas centrifuge technique is nearest to commercial exploitation. Instead of a cascade of 1500 to 2500 diffusion barriers, a gas centrifuge enrichment plant would consist of a series of rotating centrifuges. A joint project, bringing together the Federal Republic of Germany, the Netherlands and Britain, has taken this technology through to commercial development. The advantage of the gas centrifuge is its higher efficiency, both in terms of the lower energy consumption and the smaller number of stages needed to achieve the required degree of enrichment: the centrifuge's energy consumption is about a tenth that of the gaseous diffusion process, and reactor grade material can be produced by a cascade of between 10 and 30 centrifuges. All 2000 barriers in a gaseous diffusion plant are part of the basic operating unit. This means that the whole plant has to be completed before enrichment can begin. In a centrifuge plant each operating unit can be run independently, so an enrichment facility can be brought on-line in stages. The key to successful centrifuge enrichment is the centrifuge machines, and here the problem is to develop a machine that can be reliably and cheaply built – a lot of stages may not be essential for a basic operating unit, but each unit can take only a small gas flow, hence many units have to be assembled in an enrichment plant. The two international companies set up to bring this new technology into operation, Urenco and Centec, believe that they have achieved the technical basis for a successful gas centrifuge enrichment plant. Two small commercial units are being built as a part of the three-nation gas centrifuge programme. One unit, with a capacity of 200 tonne separative work units per year, is being built at Capenhurst in England; another unit, with the same capacity, is being built at Almelo in the Netherlands. The aim of the programme is to have 2000 te capacity on line by 1982, and 10000 te by 1985. (The unit of enrichment, the tonne separative work unit, often abbreviated to te by enrichment engineers, is a measure of the enrichment effort rather than the weight of product that comes out of the process. This is because different customers require different degrees of enrichment, so it is easier to classify enrichment facilities independently of the degree of enrichment of the product.) We can put this proposed enrichment capacity into perspective by looking at the enrichment requirements of a typical nuclear power station. A light water reactor with an output of 1000 MW (electrical) and a load factor of 70–75 per cent requires almost 100 te for the replacement fuel each year. Thus the 10000 te programme could support a generating capacity of 100000 MWe or more.

Enrichment R & D does not stop with the gas centrifuge. There are other possibilities that are still at a very early stage of development. Perhaps the best known of these is laser enrichment. With the massive interest and investment that is going into laser enrichment, this technique stands a good chance of succeeding beyond the laboratory stage. There are several mechanisms by which uranium enrichment might be achieved with laser light. All depend on the isotopic differences in the spectra of uranium or uranium compounds. According to Isaiah Nebenzhal of the Israel Atomic Energy Commission, 'The basic idea underlying the laser separation work is that atomic and molecular vapours absorb light only in well-defined wavelengths, specific to every species of atoms or molecules.' If it absorbs the right wavelengths of light, the 'state' of an atom or molecule can be changed. 'Atoms of different isotopes of the same element absorb slightly different wavelengths, so that each isotopic species can be manipulated independently by an appropriate choice of the laser wavelength. The same applies to molecules containing atoms of two different isotopic species of the same chemical element.' This means that 'the minute differences in properties of different isotopes of the same elements are used to bring about large changes in the state of the corresponding atoms or molecules'. As a result of these 'large changes, atoms or molecules' containing a particular isotope, uranium-235 for example, could be ionised and separated from other isotopes by electrostatic separation.

Aerodynamic separation is another possibility. In April 1975 South African scientists revealed that they had developed a novel aerodynamic enrichment system that would be the basis of a South African uranium enrichment industry. A large part of the world's known uranium reserves is in South Africa, hence that country's interest in enrichment technology. The South Africans claim that their technology is simple, and can be built up in convenient modules with a capacity between that of the gaseous

► The gas centrifuge is one way of enriching uranium – of increasing the amount of uranium-235 in the fuel. This process uses about a tenth of the amount of energy consumed by the rival gaseous diffusion process. The photograph shows that a gas centrifuge enrichment plant consists of thousands of small centrifuges. The R & D programme behind centrifuge enrichment has been concerned with developing manufacturing methods, as well as the scientific work of proving the process.

diffusion and gas centrifuge systems. Like gaseous diffusion, aero-dynamic techniques will consume significant amounts of energy; but South Africa has the benefit of inexpensive coal, which makes the energy consumption less important than it would be elsewhere.

Whichever enrichment technology is taken up, it is important to bring enough enrichment capacity into operation on time. The expense of large fuel cycle facilities is by no means negligible (see table 6.2). While electricity utilities accept that they have got to pay for power stations to be built, it is harder to find private inves-tors who will put up the money to develop and build enrichment facilities and the other fuel cycle operations. The US has had some difficulty in its attempts to 'denationalise' the country's enrich-ment business, which is the only part of the nuclear industry that has not been handed over to industrial interests outside the govern-ment. As the nuclear industry grows, so will the demand for en-riched uranium, and with it the capital needed to establish enrich-ment facilities.

The Nuclear Energy Agency's study of uranium resources, production and demand says that the demand for enrichment capacity could be as high as 124000 te/yr by 1990, with demand in 1980 likely to be between 30000 and 40000 te. According to the report, 'Existing and currently planned separative work capacity is almost certainly bound to be saturated within the next ten years. In theory, the lead time for construction of additional capacity is within that required for the nuclear power stations it will feed, but hesitations over the extension of separative work capacity could introduce significant delays.' Fortunately, the gas centrifuge should bring with it greater flexibility in enrichment plans, but thanks to the various expansions envisaged in national nuclear programmes there could still be some tight spots in en-richment supply over the next few decades.

Waste management

Short-term problems, such as the shortage of enrichment capacity, have to be set alongside the long-term doubts raised by the nature of the radioactive waste products generated in a reactor. Radio-active waste material is produced at all stages of the nuclear fuel cycle, from mining onwards. However, the most troublesome wastes are generated in the irradiated fuel elements in a reactor. The waste handling difficulties arise at the fuel reprocessing sites, rather than at the reactors themselves. At the reprocessing plant fuel elements are treated to remove the long-lived radioactive wastes. These have to be separated from the re-usable fissile materials. Each fuel element is first held for a period so that the shorter-lived radioactive isotopes can decay, leaving the longer-lived isotopes to be removed. (The 'cooling period' is normally at the reactor site, before the fuel is carried to the reprocessing facility.) Fuel elements are chopped up and dissolved in nitric acid. Plutonium and uranium are removed from the acid solution for subsequent recycling. The remaining acid wastes are the real headache. The high level wastes produced during fuel reprocessing have a relatively small volume, but they account for 99 per cent of the radioactivity in all radioactive wastes.

Just how do you ensure that something is kept totally isolated for many thousands of years? Ultimately the high level wastes, which are now held in large water-cooled tanks, will be solidified. Britain has plans to build a £20 million treatment plant that can turn radioactive wastes into glass-like solids. A demonstration plant for waste solidification should be brought into operation by British Nuclear Fuels Ltd in the 1980s. BNFL has a strong commercial incentive to develop this technology. The company makes money by reprocessing irradiated fuel from foreign reactors. Without a solidification technique, BNFL has to hang on to the highly active liquids that are produced when this waste is processed. The active material cannot be returned to the countries it came from as a liquid, but if it were solidified this might be possible. If the waste cannot be returned, Britain would be landing itself with a more serious waste handling problem than that caused by its own nuclear programme. BNFL is now negotiating 'waste return' clauses into its new reprocessing contracts.

One question that has not been satisfactorily answered is that of ultimate disposal of radioactive wastes. Solidified waste may be easier to hold than liquid waste – at least solids will not develop leaks – but it still has to be kept safe for many thousands of years. One answer might be to find some geologically stable 'graveyards' for radioactive wastes, in underground salt formations for example. This would make it difficult to retrieve the wastes if, for example, it was subsequently found that the glassy solids began to fall apart, or a better waste disposal technique turned up. For these reasons, some nuclear engineers favour storing the solidified wastes in engineered facilities where they can be monitored and retrieved if ever it became necessary.

Environmentalists do not like the idea of burying wastes. As they point out, we cannot hope to prove that radioactive wastes would remain isolated for millions of years, and that radioactive material will never reach the environment. However, nature has provided some useful evidence which suggests that geological disposal might be a truly long-term answer. It appears that millions of years ago conditions at the uranium deposits at the Oklo mine in Gabon were such that a series of 'natural' reactors was created. Scientists from the French Commissariat à l'Energie Atomique (CEA) discovered that the ratio of uranium-235 to uranium-238 in the mine was not what they expected: there should have been 0.72 per cent uranium-235 (see p. 25) but the concentration varied from place to place within the deposits. The CEA scientists deduced that this variation was due to the existence, 1700 million of years ago, of 'natural' reactors. Subsequent tests showed that the plutonium and other fission products that were created within the reactors had not spread through the underground strata. (None of the fission products remain after such a long time, but their movements can be traced by looking for the decay products from plutonium radioactivity.) Thus nature has given us pretty good evidence that it is possible to retain fission products in underground burial plots. And the Oklo wastes did not have the advantage of being sealed in strong containers.

If the thought of leaving wastes lying around for many thousands of years before they become harmless really does weigh heavily against public acceptability of nuclear power, it might be possible to develop techniques that allow some of the longer-lived isotopes to be converted into less enduring elements. There are two types of radioactive wastes: fission products, and transuranic actinides. The former are produced when uranium atoms fission: after about 500 years these have decayed to a low level of radioactivity. (Strontium-90 and caesium-137 are the most troublesome of the fission products.) The actinides are various isotopes that are produced when uranium is irradiated (but not fissioned) in a reactor core. The actinides have half-lives ranging up to millions of years. They are, therefore, nuclear power's long-term hazard, and if they could be eliminated in some way, ultimate disposal would be a less formidable problem. Actinide elements could be turned into reactor fuel so that they could be irradiated and transmuted into shorter-lived fission products. Nuclear incineration, as this technique is known, is conceptually simple; but in practice it is bound to be extremely complicated. To begin with we do not have suitable techniques for separating actinides from fission products. This is an essential part of the process: actinides and fission products will have to be separated several times if the actinides are to be completely converted. Irradiated fuel might have to be reprocessed three times to achieve 99 per cent burn-up of the actinides. Another problem would be the need to develop new fuels incorporating the actinides. And while conventional reactor fuels might achieve a 10 per cent burn-up, a nuclear incinerator should have a 90 per cent burn-up if the fuel is not to be recycled too many times. Actinide incineration might reduce the hazards to people living many thousands of years from now; but it could add to the dangers of today's nuclear workers who would have to handle a greater quantity of hazardous actinides.

The issue of nuclear waste management could be crucial to nuclear power development. The nuclear intervenors have moved away from their first anti-nuclear target, environmental pollution. Nuclear safety held the limelight for a while, but ultimately the most worrying problem could be the long-lived radioactive wastes. Waste disposal is an acknowledged problem, but those working in the business do not add to the public's confidence when they start to advocate such waste removal techniques as 'space disposal'. It has been seriously suggested that highly active radioactive wastes could be fired into orbit, or into the Sun. No matter how safe the aerospace industry may make space flight, it is difficult to see how it can convince the world that an energy system backed up by such drastic measures is worth having.

Reactor safety

The safety of nuclear reactors is another issue that troubles the nuclear industry. Most of the safety doubts have been raised for the American-designed LWRs. This is not because LWRs are inherently less safe (although the designers of competing systems sometimes hint that there may be a germ of truth in this) but because the anti-nuclear brigade has been more active in the US. One study commissioned by the AEC was the well-known Rasmussen report. Professor Norman Rasmussen of the Massachusetts Institute of Technology underlined the difficulty of making true assessments of the safety of nuclear reactors in his report (*An assessment of accident risks in the US commercial nuclear power plants*, WASH-1400). The summary report said: 'The risks had to be estimated, rather than measured, because although there are about 50 such plants now operating, there have been no nuclear accidents to date.' The Rasmussen study adapted techniques, developed as a part of the space programme and first applied to nuclear safety in Britain, 'to make a realistic estimate of these risks and to compare them with non-nuclear risks to which our society and its individuals are already exposed'. It is hardly surprising that the confirmed opponents of nuclear power found no joy in the study. WASH-1400 said that: 'The likelihood of reactor accidents is much smaller than many non-nuclear accidents examined in this study, including fires, explosions, toxic chemical releases, dam failures, air-line crashes, earthquakes, hurricanes and tornadoes, [which] are much more likely to occur and can have consequences comparable to or larger than nuclear accidents.'

Thanks to the intervenors the US nuclear industry now takes safety far more seriously than it did in the past. R&D on nuclear safety is a 'growth area' in ERDA's budget. Other issues that concern the intervenors include nuclear materials security and sabotage. What happens if a terrorist group wants to hold a country to ransom by threatening to blow up a nuclear power station? And is it possible to steal enough nuclear material to make a primitive nuclear weapon? As far as the nuclear opposition is concerned, emphasis on these topics represents another switch in tactics. Neither of these two problems is of a fundamentally technical nature. Hence the intervenors are now trying to convince the public that nuclear engineers are fallible, and that the likely consequences of any error could be so disastrous that the only way to ensure that great damage is not done is to close down the whole business. It is true that nuclear engineers cannot possibly be right all of the time; but the same is true of the aerospace engineers who design modern airliners. The question is: what are the likely consequences of any manifestation of fallibility? The opponents of nuclear power say they could be disastrous: the nuclear industry disagrees. It is almost impossible to argue this case – you either accept it or you dismiss it as nonsense. Unfortunately, the latter course is likely to reduce public watchfulness when it comes to keeping an eye on the nuclear industry. We cannot expect this industry to adopt the best path if it is not properly watched.

Nuclear process heat

The immediate problems of the nuclear industry may seem formidable enough. However, they do not deter nuclear engineers from looking ahead and developing new ideas for taking nuclear technology forward. They want to see reactors become something other than the front end of a sophisticated electricity generating system. Electricity may be a large part of a developed country's energy industry, but there is a sizeable portion of the demand for energy that cannot be met electrically. Process heat, the heat used by industry in manufacturing processes, for many applications has to be produced differently. In their present designs, reactors cannot generate heat (as hot gas or steam) at a high enough temperature for many process heat applications. The hot water from today's nuclear power stations might be fed into a district heating scheme (something that is being taken very seriously in Sweden); but the real breakthrough will come when reactors are designed specifically as heat sources. The key to this is the high temperature gas-cooled reactor (HTR). As yet the HTR is not commercially accepted and massive cost increases forced Gulf General Atomic out of the HTR power station business.

Work on the HTR concept has concentrated on the use of helium gas as a coolant. Helium is an inert gas that poses fewer corrosion problems than the carbon dioxide coolant adopted for Britain's Magnox and AGR gas-cooled reactors. With a helium coolant it should be possible to design a reactor to operate at temperatures giving a gas outlet at $750\ ^\circ C$, rising even higher if ceramic fuels are developed. The HTR is supported by sizeable R & D programmes in several countries. Europe, through the predecessor of the Nuclear Energy Agency, operated the Dragon reactor at Winfrith in Britain as a joint project. This 22 MW (thermal) reactor operated for over ten years before shutdown. The OECD energy R&D report places great emphasis on HTR development. It says that 'a marked R&D effort should be devoted to HTR reactors which might well be expected to develop substantially as from 1980'.

The generally accepted picture of the nuclear economy of, say, the year 2000 is one with just two types of reactor being built. The HTR and the breeder reactor (it might even be a gas-cooled breeder reactor) would complement each other. Breeder reactors would turn uranium-238 into a useful fuel; HTRs would consume some of that fuel and would make up the balance of the various national programmes so that the two are part of a balanced nuclear fuel system. Both would be available to generate electricity, although some HTRs would not have this as their main role. One benefit of the HTR power station would be the higher generating efficiency that comes with the higher output temperature. It might even be possible to do without the intermediate steam-raising stage. Instead, it might make more sense to recoup the energy of the hot HTR gas by expelling it through a gas turbine. The gas would be hot enough, even after it has been expanded through a gas turbine, to make further heat recovery worthwhile. This could be through a steam turbine. If high temperature heat is required rather than electrical energy, heat exchangers can extract heat from the reactor coolant gas and pass it on to a chemical processing plant, for example.

The steel industry is interested in harnessing nuclear power for steel making. The coke that is now used as a reducing agent in steel manufacture could be replaced by hydrogen made in a nuclear complex. A nuclear reactor could provide a reducing gas and heat for steel manufacture. The energy consumed by the iron and steel industry must make this sector ripe for nuclear development. In Britain, for example, it takes something like 10 per cent of the country's energy, and nearly a quarter of its solid fuel. Thus there is a clear motivation to develop nuclear steel-making methods. As a leader of the world's steel industry, Japan is in a position to appreciate the importance of nuclear steel-making, and has embarked upon a significant programme to develop production methods. And in Europe there is a nuclear steel club that brings together the interested parties in a discussion group for the development of the idea of nuclear steel making.

Fuel transformations

There have been other suggestions for harnessing nuclear power in non-electrical situations. It seems inevitable that they will become more attractive as fossil fuels dwindle and become more expensive. They will, however, call for some changes in the way that industry is organised. Reactors are built in large units, larger than many energy consumers can use. Process heat will, therefore, be supplied from a reactor to an industrial complex, rather than to individual units.

Nuclear power might provide us with a replacement for the natural gas industry, which relies on the most short-lived fossil fuel reserves. What do we do with the massive gas industry when natural gas begins to run out? Synthetic natural gas could be made from coal, or any other plentiful hydrocarbon resource (although almost certainly not from crude oil which once fuelled Britain's manufactured gas industry). Nuclear heat could help stretch fossil fuel reserves in nuclear powered gasification units. According to the General Atomic Company, nuclear energy could produce 50 per cent more synthetic pipeline gas from a given amount of coal than present fossil-fuelled coal gasification plants. Nuclear coal gasification would also be less sensitive to the price of coal than would a non-nuclear gasification plant. Doubling the cost of coal from $5 to $10 a ton for example, would increase the cost of gas by about 35 per cent in a coal-fired gasification plant but only about 22 per cent in a nuclear fuelled gasification plant. Doubling the cost of uranium would increase the cost of gas by only 2 per cent.

An alternative natural gas substitute that has many advocates is hydrogen. This gas is available in unlimited quantities if we can find an efficient way of splitting water into hydrogen and oxygen. Nuclear power could do this job. The simplest way to turn water into the two gases is in an electrolytic cell. However, electrolysis would be an expensive way of making hydrogen; thermal decomposition might be a better route. Unfortunately, no reactor could achieve the very high temperature (about 4000 °C) needed to split water in a single step. Instead it may be possible to go from water to hydrogen in a series of chemical steps that would operate at lower temperatures within reach of an HTR. The theory behind the chemistry of this process is well enough understood, but a long programme of R & D will be needed to take hydrogen production out of the laboratory and into industry.

The hydrogen economy is anything but a foregone conclusion. Indeed, there are many opponents to the idea. If we have to go to such lengths to put nuclear power to use why bother? Why not just electrify as widely as we can and throw out the gas industry when natural gas, and synthetic natural gas, run out? We would still need something to power our cars, but even there electricity has an answer. We are not far from a commercial electric car that can meet the demands of most car owners. With electricity available for many more applications, it should be possible to find, or even make, enough liquid fuels to meet essential needs, such as air travel. Just because there is a large gas distribution industry, and the technical possibility of filling the gas lines with hydrogen from nuclear reactors, this is no justification for spending valuable R & D effort.

Nuclear power may be put to many other uses, but if we judge them by their relevance rather than their intellectual appeal, many of them turn out to be of doubtful value. For example, it probably is possible to build commercially viable nuclear powered ships, but sea transport is not a large consumer of energy (when compared with steel making, for example), and should be low on the list of priority energy projects.

A question of philosophy

Today's nuclear debate is as much about philosophy as it is about technology. There are arguments as to the most 'difficult' technical problems facing nuclear engineers, but they are mostly kept within the technical community. Most of the talk is about reactors, but Norman Franklin, who was head of Britain's nuclear fuel industry until he took over the job as head of the country's nuclear reactor industry, has said that: 'The difficulties associated with reprocessing fuels of high burn-up, including the presence of plutonium of higher isotopic content, together with the need for remote operation and maintenance of complex chemical plant have presented designers and operators with one of the most challenging tasks in industrial history'. Technical headaches are mostly amenable to scientific solution, but technologists are less confident when they have to deal with the public. This may be because they are not experienced in the ways of protest and opposition; however there may also be a nagging doubt in the minds of many scientists that the anti-nuclear movement may be partly right.

One or two members of the nuclear establishment have spoken out on nuclear power's controversial issues. Chauncey Starr, president of the Electric Power Research Institute, has dismissed many of the arguments against nuclear power. For example, on the waste storage issue he has said that: 'To the professionals concerned on a daily basis with the handling of radioactive materials, this has been a "non-problem" simply because of the knowledge we have about radioactive materials, their detection and their control.' He describes the waste storage issue as 'perhaps an extreme illustration of the difficulty of bridging the communication gap between the professional and the concerned public'.

If radioactive waste management is more a philosophical issue than a technical one, reactor safety looks like a technical problem, although it too has its philosophical content. The opposition to nuclear power maintains that reactor designers cannot possibly think of all the ways in which a reactor might go wrong. However, the end point of the most serious accident at a nuclear power station would be the release of a 'substantial fraction' of the radioactivity in the reactor's core. According to Chauncey Starr, 'We do know that the only way this can happen physically is by a melting of the core fuel. Thus, the core melting becomes a key point which all initiating circumstances would have to reach before significant public hazard could occur. Thus, even if there are some unknown failure modes which might eventually lead to a core meltdown, that particular phenomenon acts as a common output for many such event sequences. Thus, if the nuclear power plant is designed to handle such an end point event as a core meltdown, the existence of a few unknown sequences presumably will not alter significantly the public hazard probabilities.'

Such arguments do not convince the nuclear opposition, which wants experimental *proof* that nuclear power stations are safe. But the Rasmussen study has shown that the likelihood of a nuclear accident is very low. According to Starr: 'the amount of experimentation that must be undertaken to establish the validity of a low probability event is so large that it would be impractical to do so. It would be fantastically expensive to run enough experiments on nuclear power stations to verify the extremely low probability of public risk, especially if some experiments involved a destruction of the nuclear core, and occasionally the containment structure. Aside from cost the time required would be equally fantastic.'

Many nuclear engineers would love to turn back to the 'good old days' when their word was accepted by the public. Science used to be viewed as something that was inherently good. People are no longer that naive. However, there must be some suspicion that some of today's anti-nuclear doubts are essentially a manifestation of a general distrust of science and technology – those 'purveyors of pollution' – and a way of challenging the wider political status quo.

Chapter 7

Fusion

In a perfect energy situation we would be able to sit back, safe in the knowledge that never again would there be any energy shortages (either real or imaginary). Some brave souls believe that one day, in the not too distant future, we could arrive at such a state of affairs. The key to this energy utopia is thermonuclear fusion, with its almost unlimited fuel reserves. Enormous technical problems stand between us and the first fusion reactor. Not least of these is the nagging fear that it may never be possible to build one. Our knowledge is not yet good enough to say that fusion power is a definite possibility.

So far fusion energy has only come to us in the Sun and the stars, and in the enormously destructive hydrogen bomb. In both these systems, and in any fusion reactor, light atomic nuclei are thrown together so violently that their enormous mutual repulsion is overcome, and the nuclei fuse to form a larger nucleus, throwing off particles carrying energy as they do so. The most promising reactions for a controlled fusion reactor involve hydrogen isotopes. Of several possibilities, the deuterium–tritium fusion reaction is most appealing:

$$D + T = n(14.1 \text{ MeV}) + {}^4\text{He}(3.5 \text{ MeV})$$

Ultimately the simple deuterium–deuterium reaction may be tapped, but this requires a higher initiation temperature.

Fusion researchers are being very careful about their 'image', despite their recent good progress. No one in the business is repeating the claims that were once made for fission. Nuclear reactors were, once upon a time, going to produce electricity so cheaply that it would be given away. Nuclear researchers earned a lot of their early R&D support with such wild claims; but fusion's budget seekers adopt another line of attack. To them it is the size of the world's fusion fuel reserves that is the real motivation for this research. (Deuterium is present in ordinary water: 1 water molecule in 5000 is 'heavy water'.) Fusion researchers are careful not to underestimate the extent of the difficulties they will have to overcome. They are not helped by an earlier embarrassing episode, which took place in the mid-1950s. At that time the world's fusion programmes were very firmly hidden behind tight security screens. These screens were torn aside with an amazing flourish of publicity. Fusion power, we were told, was just around the corner. It turned out that the scientists had misread their graphs and calculations, and that instead of eliminating the difficulties, they were discovering new ones. We are now told that these problems are slowly succumbing to the painstaking efforts of a multitude of scientists. However, the people working on fusion try very hard not to raise hopes too high. They do not want another fiasco like that of the 1950s, when their credibility took a very hard knock.

Basic conditions

Fusion can happen only if certain technical conditions are met. To begin with the nuclei have to be hot enough to overcome their natural repulsion when they collide. The threshold temperature at which fusion starts is so high that the fuel does not exist as a nicely containable gas. At 50 million degrees Kelvin (the ignition temperature for the D–T reaction) the hydrogen isotopes have lost their electrons and are a 'plasma'; that is, an ionised but electrically neutral hot gas. If this plasma touches any solid object, such as the walls of a container, it is quickly cooled and contaminated. One way of holding the D–T plasma together without letting it touch anything solid is to contain it within a magnetic field. The electrical properties of a plasma are such that a magnetic field can stop it from coming apart. This magnetic field can be created in a variety of shapes, depending upon the electrical circuitry that is used to establish the field. Unfortunately a magnetic container is not a nicely defined vessel that you can stand on end and fill with plasma. The geometry of magnetic field production is such that it is theoretically impossible to make a containment system that is completely without leaks. A hot plasma soon finds any weak spots in the field and quickly leaks out. The leaks can be through gaps in the field, or

5

through the intervention of 'instabilities' that allow the plasma to push its way out of the magnetic containment.

Twenty years of experimenting, and building ever more complex magnetic containment systems, has gradually increased the plasma confinement time, and has overcome many of the instabilities, both large and small, that cause a plasma to leak out of a magnetic field. As a result fusion scientists have devised several different types of magnetic container that might provide the basis of a fusion reactor. Some magnetic containers, such as the toroidal containment devices, are closed systems, while others are open. The tyre-shaped closed systems are more popular than the open systems. (The latter have holes through which plasma can leak.) The so-called Tokamak toroidal containment geometry has made most of the running. The magnetic containment in a Tokamak device is formed by adding two external magnetic fields together to make a helical magnetic field. Transformer coils induce an axial current flow through the plasma which acts as a secondary winding; solenoid coils around the torus produce a toroidal field that exceeds the poloidal field induced by the plasma current. The result is a plasma that is 'pinched' into a stable configuration in the middle of the torus.

The different containment geometries vary in the efficiency with which they use the magnetic field. Each plasma has its own outward pushing field and this field has to be counteracted by the magnetic containment system. The field arrangement should be such that the imposed magnetic field is not enormously higher than the internal field, otherwise the reactor's magnets are being used inefficiently and have to be bigger than in a more efficient containment geometry.

A fusion reactor has to meet some basic requirements if it is to be a net energy producer. To begin with the plasma must be hot enough; it must also be held together long enough. If these two conditions are not met the plasma will not ignite (fusion reactions will not start), and even if it does ignite a too short reaction time means that less energy comes out than goes in. The temperature at which fusion reactions start is equivalent to 4 keV (1 keV is equivalent to 1.16×10^7 degrees Kelvin), but the optimum reactor temperature will be 10–20 keV. For the D–D reaction the ignition temperature is more like 50 keV.

The 'containment time' needed to obtain a positive energy balance from a fusion reaction depends upon the density of the plasma. In a dense plasma more reactions take place in the containment period than in a more tenuous plasma. There is a theoretical expression of the relationship between plasma density, and the plasma containment time that has to be achieved in a reactor. The well-known 'Lawson criterion' says that the product of the number density (n, the number of nuclei per cubic centimetre of plasma) and the confinement time (T, in seconds) must be more than approximately 10^{14} to produce a net power output. Thus the aim of fusion research is to push nT up above this level. Over the past 20 years there has been steady progress toward this goal as new magnetic containment systems have edged nearer and nearer to the 'magic number'.

Richard Post and Fred Ribe, two workers in the US fusion programme, have estimated that in a fusion reactor, a gramme of fuel could yield 22 MWh of energy. They calculate that today's energy consumption in the US could be obtained by 'burning just 10 kg/hr of deuterium in fusion reactors'. (If this fuel really was burnt – that is, combusted with oxygen – the result would be a minuscule yield of 44 kW of energy instead of the 350 million kW obtained in a fusion reactor.) A hundred and eighty kilotonnes of coal would produce the same electrical output per hour.

Deuterium is but one of the two components of the reaction that will fuel the first fusion power stations. Tritium, the other half of the fuel cycle, occurs very little in nature (about 1 atom in 10^{17}). Tritium is a radioactive hydrogen isotope with a half-life of $12\frac{1}{2}$ years. Fusion reactors would be designed to 'breed' tritium. That is, some mechanism would be available to produce more tritium within the reactor (in the same way that the fission breeder reactor can turn uranium-238 into plutonium). The most promising tritium breeding route is through neutron reactions in lithium. The neutrons that are produced in a D–T fusion reaction have to be captured so that their kinetic energy can be turned into heat for the production of steam and electricity. If the neutrons are captured by lithium atoms, not only is their energy turned into heat, but also some of the lithium is turned into tritium through the nuclear reaction:

$$^6\text{Li} + \text{n} = {}^4\text{He} + \text{T} + 4.8 \text{ MeV}$$

So a fusion reactor could be surrounded by a blanket of molten lithium which would act as an energy-transfer medium as well as a tritium breeder.

Reactor technology

The fight for fusion power will be by no means over when the scientific difficulties have been overcome. Even when the Lawson criterion has been achieved in a magnetic confinement system, the would-be reactor builder will face a long uphill struggle. This will come as the fusion programme moves from being basically scientific into a development phase. There are some unique technological problems facing fusion engineers. They will sometimes be able to turn elsewhere for technological assistance (for example, superconducting magnets have been used in high energy physics experiments) but in other areas there is little comparable work that will help the fusion engineers.

Perhaps the most important part of a fusion reactor will be its magnets. It should be possible to achieve an energy gain of 100 to 2000 for each fusion 'event', but this significant margin could shrink to nothing if the magnets consume significant amounts of energy. Water-cooled copper magnets of the type employed in existing plasma containment devices could consume nearly all of the energy that came out of the reactor. Superconducting magnets consume less energy than conventional electromagnets to produce a magnetic field. The superconducting alloy of niobium and titanium could be developed into magnets for a fusion reactor. This alloy has been used to make large magnets in several of the world's high energy physics laboratories. The newer niobium–tin alloy could give magnetic fields twice as strong as those produced by niobium–titanium.

Superconducting technology is now at an advanced enough stage for the next generation of plasma containment machines to be built with superconducting magnets, but fusion scientists are reluctant to complicate their R&D programmes by including untried ideas in their new machines. Everything might go like clockwork, but it would be most inconvenient if the completion of an important series of experiments were held up by difficulties that had nothing to do with the experiment itself. Fusion scientists would prefer to have superconducting magnets developed as a part of a separate R&D programme.

After the difficulties with magnets, materials come next in the list of fusion technology problems. Energetic neutrons from the D–T reactions carry most of the fusion energy. As fission engineers have found, neutrons can do all manner of damage to materials. The parts of a fusion reactor will have to withstand this constant neutron bombardment. This will inevitably call for some exotic new materials. The first part of the reactor to feel the full force of the energetic neutrons is the reactor's vacuum wall. This will be bombarded by neutrons, alpha particles (charged helium nuclei), photons, and electrons. The net result of this radiation could be gradual erosion of the vacuum wall, leading to blistering and eventually to physical damage. The high wall temperature (perhaps between 500 and 1000 °C) will only exacerbate the problem. At the same time, wall material flies off into the plasma and contaminates the fuel. Further radiation effects can develop inside reactor materials. Neutron induced nuclear transitions can result in helium forming inside the material – this can make the material brittle. Other transformations are possible, and during the life of a reactor the properties of the materials subjected to intense neutron bombardment will change. Before the life of a reactor's components can be accurately estimated extensive experimental evidence will have to be gathered. Neutron sources are being developed to simulate the neutron environment of a fusion reactor.

Clearly any material that experiences such nuclear mistreatment can pose radiological headaches. To begin with there are the associated health hazards. Induced radioactivity will be a major radiological problem with fusion reactors. Secondly, heavily irradiated parts of a fusion reactor may have to be replaced during its life. One consequence of this will be the design philosophy adopted for fusion reactors: a modular approach may have to be adopted to make it easier to replace parts. For example, the first wall may have to be built in segments so that it can be replaced when radiation damage is severe.

The first wall of the vacuum chamber faces the reacting plasma and will experience the hardest bombardment by the energetic neutron flow, but other fusion reactor components will also have to withstand neutron bombardment. The superconducting magnets, for example, may have their magnetic properties modified by neutrons. One result of this may be a reduction in the current that can be carried by the material, leading to a fall in the magnetic field it can produce. Magnets will account for a large share of the cost of building a fusion reactor. If the designers have to allow for the gradual deterioration of the magnets, the economics of reactor design will suffer. They will also suffer if too much effort has to go into cooling the superconductors to near absolute zero. This extremely low temperature has to be maintained in magnets that are very near the plasma's first wall, and the hot liquid lithium.

The liquid lithium acts as both breeding blanket and coolant. Like the reactor's first wall, the liquid lithium will be at a temperature between 500 and 1000 °C. Some exotic materials may have to be used to contain the liquid lithium. Lithium is a metal and an electrical conductor, which throws up yet more difficulties. When lithium is pumped in the presence of a large magnetic field, unwanted magnetohydrodynamic effects could disturb the flow of the liquid conductor.

Plasma heaters

As if the problems of containing the plasma are not enough, fusion scientists must also find ways of putting plasma into the containment system and of heating it once it has been put there. A Tokamak's plasma is heated partly by 'ohmic' heating, in the same way that an electric heater operates: a current flows through the plasma, and the resistance of the plasma means that some energy is dumped into it by the current flow. This is not enough to raise the temperature of the plasma to thermonuclear ignition. One option for heating is to inject beams of neutral particles into the plasma. Robert Hirsch, director of the US fusion programme, describes neutral beam injectors as 'very high-current high-energy sources whose output can be efficiently neutralised, providing a high energy, high equivalent current beam of focussed neutral atoms. These energetic beams can easily cross magnetic fields and enter plasmas, where they quickly become ionised and trapped, thereby adding energy and particles to the plasma.' This technique, which is known as neutral beam injection, both heats the plasma and introduces new fuel. If the idea works, it could make it easier to achieve reactor conditions than was previously thought. Neutral beam injection can lower the target for plasma conditions set out in the Lawson criterion. According to Hirsch, it should be possible to get energy out of the plasma 'with an injected neutral beam energy of about 150 kV into a target plasma that has an electron temperature of above 5 keV and an nT value of 8–10 × 10^{12} cm^{-3} seconds'. This is about a factor of 10 below the nT required for thermonuclear fusion in a plasma without neutral beam injection.

The US will try out this new concept in plasma heating in the Tokamak Fusion Test Reactor (TFTR) which was given the go-ahead in 1975, when ERDA estimated that it might cost $215 million to build. It would have a neutral injection power of 10 MW. The US is now working on injectors capable of around 1 MW injection power. Europe's next plasma containment experiment, the Joint European Torus (JET), which was also given the go-ahead in 1975, will also try out injection beams. JET's cost has been estimated at £70 million, but this could go up as the project proceeds.

JET and TFTR will be the first fusion experiments to experience some of the problems of reactor engineering. For example, at the end of the JET series of experiments the research team may put a deuterium–tritium mixture into the machine to get some idea of how this plasma behaves. This will lead to tritium handling problems. And if the yield of neutrons in the device is appreciable, JET could begin to suffer from some of the materials problems standing between us and the first fusion reactors. Unlike a fusion reactor, neither JET nor TFTR will have the benefit of materials designed to withstand the neutrons.

Safety questions

As if getting the fusion technology right isn't enough, we must remember that in these enlightened times any developer of a new technology faces constraints that were unheard of in the past. Safety and environmental questions almost come before the 'can we do it?' question. There are four significant environmental factors: resources; tritium; radioactivity; and other pollution. The extent of fusion's *fuel* resources are not questioned, but what

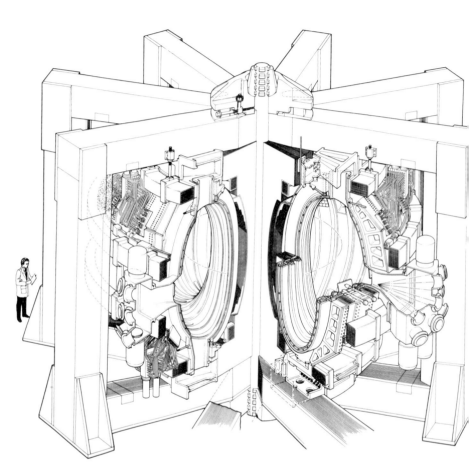

▶ Fusion reactor projects are common enough on paper, but it could be
20 years before the first true fusion reactor is built. This diagram is of
the Joint European Torus (JET) – an experimental plasma containment
device that is a part of the Common Market's fusion research programme.
JET differs from other Tokamak plasma containment systems in having
a D-shaped plasma rather than a circular plasma.

about the materials that go into the reactor? We saw in chapter 2 that lithium could be in short supply, as could the materials that go into superconducting magnets. Unfortunately, we have a meagre knowledge of these newly important resources.

Tritium, the major day-to-day safety hazard for fusion, is a radioactive material. Under normal operating conditions tritium release to the atmosphere must be kept very low. Each fusion reactor might have as much as ten kilogrammes of tritium in its fuel inventory; and about a kilogramme of tritium would be bred in the lithium blanket every day. (These are only approximate numbers as different reactor concepts postulate different quantities.) The total inventory adds up to a radiation load of 10^8 curies. The plasma in the reactor will hold just a gramme of tritium. The reactor designs that are now envisaged might release between 1 and 10 curies of tritium radiation per day.

The radioactivity generated in a fusion reactor will mostly be produced, as we have seen, by neutron bombardment of the reactor. The radioactive products will not be the same as those generated in a fission reactor. Therefore, we do not know as much as we need to about the radioactive by-products of fusion reactors. Fusion reactors will also suffer the usual environmental problems of other power stations. For example, they too will reject large quantities of waste heat.

Laser fusion

Magnetic confinement is the 'mainstream' approach to thermonuclear fusion. Another possibility has aroused significant interest over the past few years. This is laser fusion, which depends upon a completely different approach to plasma containment. With laser fusion the aim is to blast a pellet of fusion fuel – once again a D–T mixture is the likely choice – with a powerful laser beam. This beam will rapidly dump a massive amount of energy; as the laser light hits the surface of the pellet some of the material begins to vaporise and rush away from the pellet; this creates a shock wave that travels into the centre of the pellet. The shock wave forces up the density of the material in the middle of the pellet, and the conditions within the pellet can be such that fusion is sparked off. This will only happen, according to the theoreticians' calculations, if the centre of the D–T fuel pellet is compressed by a factor of at least 10 000. This system will give a positive energy output only if the initial fusion burst is followed by secondary fusion reactions brought about as the energy from the initial reactions heats the fuel pellet. This has to happen before the explosive forces sparked off in the centre of the pellet cause the whole thing to disintegrate. In the laser fusion process the plasma is confined by no more than its own inertia, and the process is sometimes known as inertial confinement.

The early days of laser fusion were obscured by a security blanket thrown over the whole business by the USAEC. This was gradually lifted, thanks partly to howls of protest from the scientific community. The declassification process continued until 1974, when the AEC allowed discussion of the design of fuel pellets being tried out in various laboratories. Soon after the topic was declassified KMS Fusion Inc. revealed that its experiments were based upon laser targets that were small glass spheres filled with a D–T mixture. Other American laboratories had adopted a similar approach in their work. The feeling within the AEC was that following this final bout of declassification, 90 per cent of the knowledge of this field had been revealed, and that anyone who wanted to study laser fusion had access to more than enough information to keep him happy.

The US has established a sizeable laser fusion programme. This received some $46 million in the fiscal year 1974, $65 million in FY 1975, and about the same in the following year. The AEC's report on the nuclear industry in 1974 said: 'Currently the largest element of the laser-fusion program is the development of high-power lasers for light-matter interaction study. Basic light–plasma interaction measurements are needed to allow advances in the theory and design of high-energy experiments which will demonstrate the feasibility of laser-initiated thermonuclear burn.' Laser fusion depends upon a sizeable R & D programme. According to the report:

'Lasers for large-scale experiments, such as significant thermonuclear burn or scientific breakeven require large-scale engineering and fabrication efforts. The development of large-scale laser systems for these experiments becomes a major program element. Two large systems are being built to study high-energy interactions with laser fuel pellets. The first, to be built at the Lawrence Livermore Laboratory, is a ten kilojoule neodymium–glass laser. Aside from being capable of being built with existing laser technology, this facility will have an experimental flexibility not possible with other lasers. The second large experimental system will employ CO_2 as the laser medium. Because of the high efficiency of CO_2 lasers, they are believed to be potentially applicable to fusion reactor systems of the future.'

The USAEC was not the only organisation to invest in laser fusion. I have already mentioned the work of KMS Fusion, a private company which had invested $20 million in laser fusion up to the beginning of 1975, when the company's founder, Keeve M. Siegel, died. This work was not the sort of thing that normally appeals to private companies, but KMSF achieved significant progress and was probably the world leader in this area of research before the big boys jumped on to the bandwagon. Outside the US, questions of 'national security' kept much of the work out of the limelight. The early leaders were in the Soviet Union, which was then followed by France. Britain took its time, and the early work was of a theoretical nature. In 1974 the UK Atomic Energy Authority and the Science Research Council tried to establish a joint laser laboratory. The UKAEA could not convince the government to pay its share of the costs so that it had to leave the SRC to go ahead with the project on its own and establish a research laboratory for the scientific study of laser–matter interactions and the development of lasers.

Laser fusion depends upon two factors: will the compression process work? and can we develop the powerful lasers that will be needed to initiate laser fusion? Today's experiments will confirm the validity of the compression theory. The work will then begin to develop efficient lasers in an attempt to produce a system that does not consume more energy than is produced by the fusion of the fuel pellet.

Research timetables

Fusion reactors will almost certainly be more complex than fission reactors. It is difficult to make any sensible estimates of the possible cost of a fusion reactor, but some 'informed guesses' put fission and fusion costs at about the same level. Before the first fusion reactor is built an extensive and expensive R&D programme will have to be conducted. Plasma containment machines have yet to reach reactor size and conditions, and they are already extremely expensive. Fusion research is one of the few areas where there has been good international collaboration between 'competing' countries. The Soviet Union, the US, and Europe each have significant fusion R&D programmes. Like other energy budgets, the US spending on fusion has soared over recent years. It rose from $53 million in FY 1974 to $85 million in FY 1975, with the budget for FY 1976 somewhere near $120 million. Europe's fusion R&D comes under Euratom's umbrella. Until recently this was mostly a case of funding a part of the various national programmes, but Euratom's fusion work has now come to a stage where it must establish a joint research project to assemble and operate the Joint European Torus (see p. 126).

Just when the various fusion programmes might come to something is anybody's guess, and several people in the business have put forward their guesses of when we might see the first fusion reactor. Perhaps the most detailed analysis of the prospects of fusion is that prepared by Dr Robert Hirsch, head of the US fusion programme until 1976. Early in 1974, he said:

'In recent years the controlled thermonuclear research program has experienced a number of major successes wherein significant experimental results were achieved and these were as predicted by preoperational theory. Further, the outlook for continued success is considered excellent; remaining problems are believed to be well posed and soluble. As a result, the program recently developed a detailed long range plan, which includes a number of major goals. These recognize the physics problems that remain to be addressed as well as the arduous engineering development that is an inherent part of the commercialization of any new prime energy source.'

Hirsch's timetable for fusion power breaks down the programme into readily identifiable steps. To begin with, machines now being built will achieve conditions approaching those of a reactor. At first these plasma conditions will be achieved with hydrogen plasmas rather than a genuine fusion fuel. The second step will be to achieve reactor plasma conditions with a D–T plasma: for the first time this will yield fusion reactions under reactor conditions. The next step, the third in the future development programme, is an experimental power reactor with a small power output. This will be followed by a larger experimental power reactor. At the end of this R&D chain, Hirsch tells us, the demonstration power reactor will be built to operate sometime around the year 2000. Hirsch says, 'DT burning systems are expected to operate about 1980 and for the first time produce significant fusion energy for strictly peaceful purposes.' The first experimental power generating fusion reactor might start working in the latter half of the 1980s.

If the first industrial fusion power station starts working early in the 21st century, we can expect fusion generating capacity to grow gradually until it can take over the role of the major electricity generator. By then the established electricity supply system will be that much larger than it is now; thus it will be some time before fusion is providing a significant fraction of our energy.

Further ahead

A few brave souls have tried to look beyond the fusion power station in search of possible non-electrical uses of fusion power. Fission technology is only now being taken into unknown territory, so it may seem somewhat premature to start talking of fusion in such terms. James Powell, of the Brookhaven National Laboratory, does not believe that it is too early to think about other applications of fusion. He has studied the production of synthetic fuels by fusion reactors and maintains that, given the anticipated improvements in process efficiency, fusion reactors could produce hydrogen as cheaply as gas from coal. Fusion reactors may be preferable to fission reactors for the production of synthetic fuels. KMS Fusion has even worked on a laser fusion technique for the production of hydrogen by a complex cycle of 'chemical nuclear reactions'.

The fusion 'torch' is another idea. This is a sort of nuclear garbage disposal unit. The extremely hot plasma from a fusion device would be turned into a beam. If material, urban waste for example, is fed into the plasma beam it can be heated to such a high temperature that its atoms dissociate and ionise. The different elements in the waste can then be separated from one another. William Gough and B. C. Eastlund came up with this idea. They have said of the fusion torch:

'Waste products could be converted back to elemental form, thus simultaneously eliminating the problem of the disposal of the waste product while providing a supply of basic raw materials. When fusion reactors are available...thousands of tons per day [of material] could be handled in a single plant. The handling of urban wastes then becomes a possibility. If just 10,000 MW were used for processing via the fusion torch, the entire wastes of a city with a population in the millions could be converted back into raw materials each day.'

A group at the Battelle Pacific Northwest Laboratories has looked into the possibility of transmuting high level radioactive wastes from a fission reactor. The group concludes that:

'Within the limits of the data and techniques used in [our] analysis, CTR [fusion] transmutation of high level wastes is feasible and attractive, although it is more attractive for the actinides than the fission products. This is fortunate in that the actinides constitute the greater long term hazard. It therefore appears that CTRs can potentially transmute most of the high level wastes from a fission economy into stable or short half-lived ash.'

It would be a strange quirk of fate if, when they appear on the scene, fusion reactors are given the job of clearing up the mess left by fission power.

Chapter 8

Solar power

The Sun's free, limitless, and non-polluting supply of 'fuel' has great emotional appeal, particularly to those who see nothing but damage when they look at conventional energy sources. More sceptical souls see solar energy in purely economic terms: it will be introduced, they say, as and when it makes financial sense. The emotional appeal of solar energy – no one is actually *against* the idea – has encouraged the growth of a small but expanding market for solar hot water systems. That some of these systems are, to put it mildly, of dubious merit makes life difficult for the genuine solar energy industry.

Solar energy systems are not new. Earlier this century they were very popular in Florida, for example, where solar hot water systems sold well until other fuels became so cheap that solar energy systems lost their appeal. Even today there is a sizeable solar energy industry. A report prepared in the US by the National Science Foundation and the National Aeronautics and Space Administration said, in 1972, that 'solar water heaters are commercially manufactured in Australia, Israel, Japan, USSR, and on a small scale in the US. The aggregate business of these enterprises is probably several million dollars per year.' A report from the Australian Academy of Sciences said of the local solar heater industry: 'There is already a small, efficient, solar water heating industry, backed by local research and development, which produces general purpose flat-plate absorbers.' At that time, in 1973, the industry was producing about 4000 square metres of collector each year.

Today's renewed interest in solar energy stems partly from a belief that the Sun can compete on economic terms with oil and gas. There is no doubt that 'free' solar radiation is cheaper than other fuels, whatever their price, but the capital cost of a solar powered hot water system, for example, is, with today's technology, usually higher than the cost of, say, a gas or electric hot water system. This higher capital investment makes sense only when the lower running costs of a solar energy system lead to high enough savings to recoup the higher capital cost over a reasonable period. Such a state of affairs exists in remote parts of Australia, for example, where the cost of carrying oil to the consumer is so high that solar energy systems already are economically viable.

Beyond this strictly economic view of life there is another motivation for reviewing the potential of solar energy. That solar power might be a financially viable alternative to fossil-fuelled became clear when the price of oil went up; that it might have greater significance has been realised since people started to worry seriously about the ultimate 'death' of fossil fuels. No matter how much we turn coal into oil and gas, fossil fuels will run out. In fact, by making coal a universally applicable fuel we are hastening the day when coal too will join natural gas and crude oil as a thing of the past. Coal may not run out for several centuries, and there is always nuclear power to help stretch our energy reserves. If fusion power proves possible we may not have to worry about energy reserves ever again. But we cannot be absolutely sure that fusion can be deployed to provide all the energy we need. Fusion technology still is well and truly laboratory bound. And if ever the fears of the foes of fission are realised and the world witnesses a truly disastrous accident at a nuclear power station, the growth of nuclear power might be brought to an abrupt stop. This would put even greater stress on the world's fossil fuels. Solar energy is just about the only energy system that could conceivably step in if fission comes unstuck and fusion proves to be untameable. Thus, by studying solar energy and developing solar power systems, we are investing in an insurance policy that may never prove necessary. However, if solar energy is developed to an advanced stage it is most unlikely that it will be ditched even if fusion power is a success. A more likely scenario is one in which the two energy systems eventually take over from other energy systems, and end up doing those jobs that they are best suited to.

Uncertainty about the possible future prices of fossil fuels must add another motivation for the development of solar power. Fuel prices will rise in the long term as first gas and oil, and then coal, begin to run out. We should not expect this particular price mechanism to come into effect in the near future; however, thanks to the massive price rises over recent years – and these hit all fuel systems – hedging against possible future price rises is now a popular pastime. Hence the conventional economic ground rules have changed; people now seem to be more ready to make larger capital investments if they anticipate that future fuel prices might make it possible to recoup them more quickly than they can at current prices.

The flow of energy

The amount of energy that comes to us from the Sun was outlined in chapter 2. The World Energy Conference survey of energy resources says that the solar radiation that arrives at the Earth's surface is something like 20000 times as much as the present world production and consumption of 'commercial energy' from all sources. So there is plenty of solar energy available. This radiation is not intense, it isn't there at night, and when the weather is poor the Sun's power is drastically diminished. Sunlight's unpredictability makes it less attractive in some areas than in others. Countries such as Britain are in a sort of twilight region, where solar energy is neither definitely attractive nor positively irrelevant. Unfortunately, many of the countries hardest hit by the energy turmoil of recent years not only suffer from a shortage of energy reserves, but also fare badly in the solar energy stakes. Europe has only a limited solar energy potential. With its wide climatic variations the US is much more suitably placed to implement solar energy systems in its sunnier regions. In its influential report on the prospects for solar energy in the US a joint NSF/NASA study group said that the average yearly incidence of solar energy on the ground in continental US is 17 watts/sq.ft. This results in 'an average daily (24 hour) energy supply of 410 thermal watt-hours/sq.ft. This value is approximately twice the amount needed to heat and cool an average house.'

While solar energy can, theoretically, be harnessed for many energy tasks – direct generation of electricity is just one proven possibility – solar power will probably have its first impact as a domestic heat source. Large central power stations may be developed, with large heat generating machines or direct conversion into electricity, but the motivation to centralise in this way is questionable when solar energy's great benefit is that it is delivered everywhere, or at least everywhere that might expect to be able to use it. When this is the case, why develop centralised systems that harness the energy at a central location and then distribute it through a transmission system? Is it not more sensible to decentralise the whole process? This is just one argument that is exercising solar energy researchers.

Solar energy makes sense for domestic use because domestic consumption is an appreciable part of total energy consumption (between a fifth and a quarter). And space heating makes up by far the largest part of domestic energy needs (about 70 per cent of the energy that goes into British homes). Add to this the energy that goes to heat water and that needed for air conditioning in those countries where it is common, and it becomes clear that domestic heating and cooling add up to one of the largest uses of energy. The Sun can provide this energy more easily than it can meet other energy requirements.

The potential contribution solar energy can make to a country's energy needs clearly depends on geographical factors. The various reports on solar energy come up with some estimates of the possible penetration into the energy 'market' that solar power might make in different countries. The NSF/NASA report concludes that 'a substantial development program can achieve the necessary technical and economic objectives by the year 2020. Then solar energy could economically provide up to 35 per cent of the total building heating and cooling load; 30 per cent of the Nation's gaseous fuel; 10 per cent of the liquid fuel; and 20 per cent of the electric energy requirements.' Another conclusion is that 'if solar development programs are successful, building heating could reach public use within 5 years, building cooling in 6 to 10 years, synthetic fuels from organic materials in 5 to 8 years, and electricity production in 10 to 15 years'. Since this report was published (in December 1972) the oil upheavals of 1973/4 have altered the basic framework, and solar energy could make a more rapid penetration of the market, although the report's optimistic estimates of the possible contribution of solar energy to fluid fuel markets are not supported by less partisan observers. At a symposium held in April 1974, Lloyd O. Herwig, who was then director of the NSF's advanced solar energy research and technology work, talked of the six areas of solar energy research: heating and cooling of buildings; solar thermal energy conversion; photovoltaic conversion; wind energy conversion; biomass production and conversion; and ocean thermal energy conversion. Herwig maintained that 'no technological breakthroughs are required to obtain useful energy and power from early conceptualizations of these solar energy systems'. (Herwig's definition of solar energy is wider than that covered in this chapter: some of the subjects in his list are left for discussion in the next chapter.)

An Australian report makes some even more optimistic projections. The report of the committee on solar energy research in Australia was published in September 1973 by the Australian Academy of Sciences. This suggests that 'a target of 2×10^{18} joules per annum be set for the solar contribution to low grade heat and transportation needs in the year 2000, by which time it is estimated that the nation's energy needs will be 7.5×10^{18} joules per annum'. Here too the subsequent rise in oil prices must throw doubt on the projection of future energy needs – most countries have marked down their estimates of future energy consumption – but this can only increase the Sun's likely share of Australia's energy market.

Britain and other European countries cannot expect such dramatic growth in solar energy utilisation. A study conducted by the Central Electricity Generating Board (which would not deny its partisan viewpoint, but might try to disguise it by saying that it is more objective than solar energy enthusiasts) concluded that in the year 2000 Britain might expect solar energy to contribute an amount of energy that corresponds to 4 per cent of *current* total demand. The CEGB study points out the inertia that is built into the system: 'If all new houses built from now use only solar energy for heating water, then in 30 years' time, solar energy would be contributing an amount of energy equivalent to about 2 per cent of current total energy demand, that is about 8 per cent of current domestic energy demand'. And on a global basis 'the effect of solar energy upon the supply and hence market price of the world's fuel will be rather marginal'.

Solar collectors

The technology of solar energy is far from new. Some of the experiments with solar energy are recorded in engravings rather than photographs. Sun-powered engines, water distillation units, solar water heaters, a gigantic solar furnace, crop dryers, food cookers, complex solar cell arrays for space craft, and 'educational' toys have all been built at one time or another. Some have even been exploited commercially, as we saw earlier in this chapter. The US National Academy of Sciences detailed some of the experience with solar energy systems in its report *Solar energy in developing countries: perspectives and prospects* (published in March 1972). This said that solar water heaters are:

'proving to be the most economical way of providing hot water in many parts of the Australian continent, particularly in the Northern Territory, where it is Australian government policy to install solar water-heaters in all government houses. Hundreds of thousands of water heaters have been manufactured in Japan...Experiments with solar water-heaters have been undertaken in many areas, such as Chile, India, Egypt, etc...However, use in countries other than the United States, Israel, Australia, and Japan has, to date, been quite limited.'

And the CEGB study concluded that even in Britain 'it appears just about economic, at least in the south of the country, to heat water by solar energy'.

Flat-plate solar collectors are simplicity itself. The collector is just a flat black plate. (Copper is the most commonly used material but aluminium is making a bid for the job.) In most collectors there is at least one glass plate in front of the flat plate to reduce the heat lost by re-radiation. The heat that builds up in the collector is removed by water running through pipes either built into the metal collector or running over its surface. All manner of changes can be worked on this basic idea, and the aim of any research programme must be to arrive at the most efficient design. 'Most efficient' does not mean making the collector that produces the highest heat output from a given surface area. The fuel is free, so the aim is to reduce the capital cost per unit of energy output. It may be that a cheaper system can be achieved with a lower solar energy collection efficiency. In this way it is possible to reduce the capital cost of the system, a key factor in determining the acceptability of solar energy systems. The task of solar energy R & D is not finished when a reliable and inexpensive collector comes on to the market. It is also important to develop the peripheral hardware that makes solar heaters efficient in operation. This means studying the whole system. Many researchers looking into solar energy for the first time go through the process of redesigning the collectors themselves from scratch. This is a harmless but time consuming exercise. The more important research that is underway in this area is on the application of solar energy in buildings.

'Solar houses', as most experimental buildings to harness solar energy for domestic heating and cooling are labelled, are far from new. There are and have been plenty of them built from time to time. Whereas in the past they have been built, and mostly lived in, by dedicated believers in the cause of solar energy, more and more experiments are being supported on an 'institutional' basis. One of the early solar-house experiments was set up at the Massachusetts Institute of Technology (MIT) in 1939. Like most subsequent buildings, MIT Solar House 1 (it was followed by numbers 2, 3 and 4) had a heat store. The heat store in such houses was often a large tank of water (MIT 1 had a 17 400 gallon water tank). Sometimes the heat store has been no more than the thick foundations the house was built on. Norman Saunders used this system in a solar house he built in 1960 in Weston, Massachusetts. This house's solar energy system provided 60 per cent of the winter heat demand. A recent survey of solar heated buildings listed more than 50 houses that had been built and operated at one time in the US.

Solar collectors are much loved by the commune movements. The 'commune-ists' move into their farmhouse and immediately think of the sort of non-polluting, resource-miserly systems that they can put into their new home before they pull out the electricity cables and tell the coal man to stop deliveries. Most solar roofs can be made to work; but though they at one time made up much of our experience with solar energy in housing, such one-off experiments do little more than tell us that solar power works. It is not the one-off farmhouse conversion that will help find solutions to our energy problems.

▶ There are plenty of experimental solar energy systems in the world, particularly in the US. Honeywell installed this solar heating system at a 1300-pupil school in Minneapolis. A solution of water and glycol circulates through 246 collectors, and is heated to between 55 and 65 °C by the Sun. The heat warms the school building and provides hot water.

The efforts to develop systems that might be adopted in commercial buildings or other large buildings such as schools or offices are far more relevant. So too are the housing projects that involve solar energy application to dwellings that are more representative of the houses in which most of us live. When NASA started building a solar heated single storey building with solar heating and cooling, it described its Langley project as 'the first building of its size in the world for which solar energy will provide a significant part of the building's heating and cooling load'. The 53 000 sq.ft building is a systems engineering building, with a 15 000 sq.ft solar collector. The associated R&D programme has tested solar collectors in a 'solar simulator producing typical sunshine conditions'. It may seem rather a waste of time to build a fake Sun when the real thing is available, but the NASA simulator can duplicate conditions ranging from 'a cloudy day in Cleveland to a cloudless one in Arizona'. (Langley is in Virginia.) At the time it started work on the project, NASA estimated that the collector should cost between $1 and $2 a square foot if the system were to be widely adopted. NASA maintained that the key to the acceptability of solar collectors was their cost.

Early in 1974 the US National Science Foundation proudly announced that 'a typical elementary school' had become 'the first in the nation to obtain heat from solar energy'. It cost more than $$\frac{1}{2}$$ million to build a 5700 sq.ft collector system for the 9581 sq.ft of the wing used in the experiment. The solar collectors were put on the roof of the building. Hot water room heating was used, in conjunction with a 15 000 gallon storage tank. The school was the Timonium Elementary School, near Baltimore, Maryland.

These were just two experiments in a rapidly growing list. The NSF played a major part in this programme until the Energy Research and Development Administration took over the solar energy programme. One project carried out with NSF funds, and some support from Honeywell Inc., Minneapolis, Minnesota, was the wandering laboratory. It is not too difficult to make quick calculations that give a rough estimate of how much energy the Sun sends down to a particular area; but there is no real substitute for a detailed programme of measurements. What better way to do this than to tow a mobile laboratory from place to place, making measurements as it goes? Honeywell's Systems Research Center designed and built the laboratory, which is housed in two large trailers. One trailer carries the solar collector system (650 sq.ft of collector panels covered by transparent plastic and glass), and the measuring instruments; the second trailer carries the heat 'load' (a small office that is heated, cooled, and supplied with hot water from the solar energy system). The mobile laboratory has been driven around the US on visits to some of the sites of solar energy projects – one of its stops was the Timonium Elementary School.

The enormous potential of 'solar climate control' – the use of solar energy for space heating and air conditioning – was described in a report prepared by Arthur D. Little Inc., the international research and market research organisation. ADL estimated in October 1974 that 'the market for solar climate control could reach \$1.3 billion by 1985 in the US alone, if industry, with effective government support, moves ahead promptly to introduce solar hardware into the marketplace'. The staggering potential of solar power was put at such that 'by the year 2000 annual energy savings could be 2 million barrels of oil daily'. ADL did not rule out the use of solar heating in Europe's less friendly climate: 'Cost projections developed for a number of European cities indicate that solar heating is economically feasible in many locations when compared with systems using conventional fuels.'

Britain's solar outlook cannot hope to match that of the US and other, sunnier countries; but Britons should not rule out sunpower simply because they love to talk about their miserable weather. In a report prepared for a parliamentary committee that was looking into the country's energy resources, the UK branch of the International Solar Energy Society put the issue into perspective: 'The UK climate is not particularly favourable for solar energy utilisation if only direct sunshine is used. However, if proper use is made of both direct and diffuse energy, the prospects are much better. Differences between very hot countries and the UK are much smaller than most people think.' The report goes on to say that 'there are considerable possibilities for making more effective use of solar radiation in buildings by adopting appropriate architectural designs'. In other words, when designing buildings architects should 'think solar' and look for ways of allowing the Sun to add its bit to the energy input of the building.

There are various experiments under way in Britain to try out this approach to solar energy utilisation. At the new town of Milton Keynes the Department of the Environment, the government department responsible for housing, has funded a pilot project to see if solar heat can be harnessed in a typical house. Solar energy should provide about 60 per cent of the heat energy needed by the occupants of this three-bedroomed terrace house. In another experiment, near Liverpool, a housing development has conventionally heated dwellings alongside solar heated houses. This means that the two can be directly compared. The nine solar heated houses are built with high density brick, with one outer wall covered by a double-glazed skin. The wall absorbs solar radiation and re-radiates it into the house. The wall is, in effect, a giant storage heater that gets its energy from the Sun. This concept was pioneered in France, where there has been talk of building 200000 to 300000 solar houses a year. In one French experiment, at Odeillo in the eastern Pyrenees, cuts in energy consumption of 70 per cent have been achieved in the heating system. And the solar energy system added only 8 per cent to the cost of building the experimental dwellings.

In the projects outlined here, the Sun is a source of heat energy. There are other, more ambitious solar dwelling developments under consideration. One concept -- known as the solar community – comes from the Sandia Laboratories, Albuquerque, New Mexico. Sandia says its proposal 'envisions the central collection and storage of solar energy and its distribution to individual homes and businesses to meet the need for electricity, heating, air conditioning, and hot water'. The Sun would do just about everything, but there would not be a large central power station with a sizeable distribution network. The Sandia concept could give the best of both worlds: a larger system than the individual house-sized unit would be cheaper to build, but the complex would be small enough for the solar energy to be collected locally. Sandia says its concept 'derives its economies from systems analysis which has helped to maximise the contribution of each element in the system'.

A solar community would have centralised collectors to provide hot water which would be stored underground in insulated tanks. This heat store would drive turboelectric generators, provide

► The solar furnace at Odeillo in France is a dramatic demonstration of the power of the Sun and one of the best known solar energy projects. In fact the furnace is not so much an experiment to develop a new energy system as a materials research project. Solar energy can be focussed by the array of mirrors to provide a temperature over 3000 °C without introducing any fuel impurities or contact with the heating system. The power level of the solar furnace can rise to 1 megawatt. The Sun's light is reflected on to the mirrors on the curved focussing surface by flat mirrors on a hill in front of the giant collector.

energy for absorption air conditioners, and meet space heating and hot water needs. A solar community might have 100 to 1000 units (houses, flats, small buildings); and the Sun would meet perhaps 60 per cent of the energy needs – a fossil-fuelled backup would be cheaper to install than extra solar energy storage capacity to meet occasional shortages.

A solar community might cost 20 to 100 per cent more to build than a purely fossil-fuelled community, but, according to Sandia, 'The project's costs are estimated to be less than the cost of producing electricity from central power plants because both electric and thermal energy are used. Costs would also be less than those for systems for individual dwellings since the community use of equipment permit the cost of this hardware to be divided among the many units to which energy is supplied.'

Clearly there are many ways in which solar energy can be put to use. This underlines the importance of the 'systems' approach to solar R & D. The research question is not 'Can we do it?' but 'What is the best way to put together available technology to meet known needs?'

Solar cells

Electricity generation is the aim of many energy R & D projects: solar research is no exception. There is no shortage of ideas as to how the Sun might yield electricity, indeed, solar cells represented the largest market for solar energy equipment throughout the 1960s. However, existing technology is not designed to meet everyday needs. Solar cells can turn sunlight directly into electricity. They were little more than a curiosity, and a convenient light-measuring device, before the space race suddenly threw up a requirement for power systems that could provide electricity for satellites and other space systems. Perhaps the most spectacular orbiting solar energy system was the Skylab 'windmill', with its four large 'sails' of solar cells and two supplementary cell panels. (One of the panels caused a lot of headaches when it failed to deploy properly and threatened to wreck the whole mission.) Unfortunately the price that NASA can afford to pay for solar cells in prestigious space projects far exceeds that which would be acceptable at ground level.

The requirements for space solar cells are not the same as for terrestrial applications. Beyond the irrelevance of price for space projects, weight is more important in space so the aim was to build cell systems that provided the greatest amount of electricity from the lightest power units. And in space you cannot send an engineer along to put things right if a lead comes unsoldered, for example, so reliability is very important. On Earth sunlight is a free fuel, which makes it more important to produce cells with the best power output for a particular capital investment rather than the lightest system. As things now stand solar cells cost something like fifty times too much to make them viable for terrestrial uses. Even this is a step forward in comparison with the cost a few years ago. Whereas in 1975 you could buy solar cells for about $30 per peak watt of energy (the output when the sun is at its brightest), the price was nearer $100 four years ago. The price has to come down to about $1 or less before solar cells become commercially acceptable for anything other than specialist applications.

These cost reductions will come about only if there is significant progress in the manufacturing technologies used to produce solar cells. Mass production will cut the cost of cells. Other necessary advances include cheaper silicon materials, which might require the development of cells based upon poorer quality silicon, and improved cell fabrication techniques, including new ways of making large silicon slabs and automatic cell assembly. One new silicon production technique has been developed by Tyco Laboratories, Waltham, Massachusetts. Whereas cells have traditionally been made by growing a high quality crystal of cell material and then cutting it into thin layers for subsequent mounting in power units, Tyco's technique produces a continuous ribbon of silicon solar cell material.

It seems inevitable that a major part of the cost of a solar electric system will be the price of the storage batteries that will be needed to provide power when there is no sunlight. There is at least one possible way round the energy storage problem for solar electricity generation. There are places where a solar collector can be placed so as to maintain an almost continuous energy flow – an orbiting solar cell panel in synchronous orbit, where the satellite stays over the same place on Earth, could receive uninterrupted sunlight most of the time. The space solar power idea is the brain-child of Peter Glaser of Arthur D. Little Inc. Glaser describes his idea as two symmetrically arranged solar collectors which convert solar energy directly to electricity. This power is fed to microwave generators built into the satellite's transmitting aerials. These direct the microwave beam to a receiver on Earth where the micro-wave radiation is converted back to electricity. Not only does this system eliminate daily blackouts and bad weather periods, it also sees brighter sunlight. (Half of the light from the Sun bounces off the atmosphere back into space.) Glaser has made some calcula-tions of the costs involved in such a system. In 1973 he estimated that the capital costs of a prototype system designed to generate 5000 MWe on Earth would be $1040/kWe. A third of this cost would be for the solar cells and solar collector array; a quarter of the cost would be the microwave generators, transmitters, re-ceivers, and so on; finally, the cost of putting the system into space might be $500/kWe. When Glaser presented this idea to an inter-national meeting on solar energy there was some opposition to the 'high technology' nature of the idea. Glaser himself believes that less advanced technological systems will have a major part to play in the development of solar energy; but he believes that if we wanted to develop the orbiting solar cell system it might take 20 years or so, which compares well with the time it takes to de-velop something like thermonuclear fusion, or even the breeder reactor.

Power stations

Solar cells are not the only way to turn sunlight into electricity. Sandia's solar community would generate electricity with a modi-fied generator cycle (see p. 143). And there are other potential solar–electricity technologies. There is even the possibility of an equiva-lent of today's massive central power stations. Efficient and long lasting solar energy concentrators can produce a high temperature to generate steam or some other working vapour for turbine operation. The working fluid in a conventional turbine cycle has to be at a much higher temperature than anything that can be reached with the flat plate collectors that are suitable for domestic heat supply. The higher temperatures can only be reached if the solar collector focusses the light from a large area on to a smaller area. And focussing collectors have to be steered if they are not going to miss much of the energy that comes from the Sun. A flat plate collector can achieve temperatures up to about 150 °C. Focussing collectors that can be steered to follow the Sun can produce temperatures from hundreds to thousands of degrees Centigrade. One of the best known research teams working on this branch of solar energy, the group led by Aden and Marjorie Meinels at the University of Arizona, selected 1000 °F (540 °C) as their operating temperature. If the temperature of the working fluid is to be over 500 °C, the collectors must trap the heat and re-radiate as little as possible. This calls for 'selective coatings' on the collector surfaces. These coatings are transparent to sunlight, but are opaque to the infrared radiation that would take the heat energy away from the hot fluids. As the Meinels put it: 'We want a surface that is "black" to sunlight but looks like a perfect mirror in the infrared.' And these coatings have to be inexpensive and long lasting, and amenable to manufacture in a factory. There is no point in inventing the world's best selective coating if it is extremely expensive and can be made only in a sophisticated research laboratory.

Most of the proposed designs for large solar power stations would depend solely on the Sun for their energy input. Another intriguing possibility is a dual solar/fossil fuel unit. This idea has been put forward in the US by engineers at the Foster Wheeler Corporation. In a 'hybrid' power station solar energy would supplement fossil energy, so that when the Sun shines brightly enough the station would reduce its consumption of fossil fuel. Thus there would not have to be a solar energy storage system to combat dull weather. The economics of the system would, however, be complicated by the two-fuel arrangement. The fossil fuel saved by solar energy would have to justify the cost of the solar system plus the non-fuel costs associated with that portion of the fossil unit that sits idle when the Sun is shining. According to Foster Wheeler, when capital cost, energy conversion efficiency, and technological complexity are taken into account, the most promising option is to employ solar energy to superheat steam that has been raised by burning fossil fuel. The company has designed a steam superheater that looks a bit like a lighthouse, with the solar collectors where the light would be. Steam runs up and down the tower and is heated by the solar collector. The engineers working on this project reckon that practical limitations would restrict the solar energy input to about 20 to 50 per cent of the power station's total energy consumption.

A large power station based on solar energy would cover a sizeable area; for example, in the south western United States (one of the more popular areas for a solar power station) something like 10 square miles (25 sq.km) of solar collectors could 'fuel' a 1000 MWe power station. As far as technology goes, a solar power station would not require great advances. As the NASA/NSF report said: 'There are no technical limitations that would prevent a solar thermal power station from being built today. The question is whether it would be economically competitive with other methods of power generation. It is this question which has to be answered to make the approach a reality.'

Other solar energy ideas could produce energy without taking up large areas of land. The ocean surfaces between the tropics of Cancer and Capricorn stay at a temperature which never falls below 28 °C (82 °F), thanks to the constant flow of solar energy. Some way below this warm surface sit the cooler waters that have come from polar regions. 'It is thus possible', says the NASA/NSF study, 'under several hundred million square miles of ocean to find a nearly infinite heat sink at 35 to 38 °F, at a level as little as 2000 feet directly beneath a nearly infinite surface heat reservoir at 82 to 85 °F. Both heat reservoir and heat sink are replenished annually by solar energy. A heat engine operating across a 50 °F temperature difference in an 85 °F heat source would be able, theoretically, to convert to useful work 9 per cent of the heat flowing across it.'

Dixy Lee Ray's WASH-1281 report included ocean thermal gradients, as this concept is known, in its recommended programme for energy R&D in the period 1975–80. The report said that the US should 'determine the technical feasibility of producing electric power from ocean thermal gradients by laboratory-scale testing of prototypes and full-scale testing of necessary components'. It said, further on, that the key elements that have to be studied are 'the heat exchanger, the deep-water pipe, and the overall plant structural facility'. WASH-1281 suggests a five-year budget adding up to a total $26.6 million. According to Professor Clarence Zener of the Carnegie–Mellon University, who has studied solar sea power, 'an optimally designed plant need cost no more than a conventional fossil-fuelled plant'. Zener estimates that a square mile of ocean could yield a megawatt of electrical energy.

Fluid fuels

Thermal and electrical energy have to be backed up by a supply of fluid fuels. Here too solar energy is looked at as a possible source. Biological processes offer one way of making solar 'fuels'. Photosynthesis is, however, far from efficient: only a few per cent of the incident radiation is converted into plant growth. And, while trees may have been one of the first fuels, it is not efficient to burn most plant material without first drying it, which itself consumes energy. Gas and liquid fuels, however, can be made from plant matter in several ways. Biological material of all sorts can be fermented with reasonable efficiency. And plant material can, like coal, be converted to liquid fuels by pyrolysis.

The NASA/NSF study on solar energy says: 'If, through advanced management practices including exploitation of modern developments in plant genetics, plantations were operated to produce continuous crops at greater than 3 per cent solar energy conversion, less than 3 per cent of the land area of the US would produce stored solar energy equivalent to the anticipated US electric energy requirements for 1985.' The numbers are encouraging enough for a country with as much land as the US; but in Europe it would be unthinkable to grow plants for energy on such a scale. Even in the US, energy plantations would have to compete with food plantations; and in these days of food shortages, when millions of people in the world are slowly starving to death, it would be a brave man who would allocate valuable agricultural land to meet the energy needs of a profligate society.

Research growth

The OECD report on energy R & D says that: 'Research into solar energy is relatively advanced as compared with that on all sources other than conventional and nuclear ones.' However, it will be some time before solar energy begins to make inroads into the existing energy market. Small-scale applications of solar energy, to heat domestic water and buildings, for example, may be the most advanced systems, but 'allowing for social–economic factors such as consumers' reluctance to accept higher initial construction costs despite lower operating costs, and the slow rate of new house building, much more than a decade will elapse before this form of heating can come into a more general use and affect total energy consumption'. The OECD report is less optimistic about large-scale applications such as solar power stations:

'The economics of a solar power station, given the very high capital costs involved, are by no means comparable with those of conventional thermal or nuclear plants. Estimates indicate that solar power plants would be at least five to ten times more costly than conventional plants in view of the very high maintenance costs. Thus for a 1000 MW peak capacity thermal power plant the investment required would be 2000 to 2500 million dollars and a mean kilowatt of capacity would cost $2500.'

Research and development have to be backed up by the right institutional structure if solar energy, or any other new energy technology for that matter, is to be adopted. The report *Canada's energy opportunities* detailed this aspect of energy development: 'Solar energy will only become a commercial reality when we have concentrated research and development effort in conjunction with industrial entrepreneurship.' And if the first impact of solar energy is to be for space heating buildings, 'the housing industry must be directly involved in order that the new techniques not remain unapplied in the design shops, nor be restricted by outdated building codes and regulations'.

Industry is showing some interest in solar energy. In the US, for example, the oil companies, which are fast turning themselves into energy conglomerates, are spending money on solar energy R&D. However, most of the spending is done by governments. The OECD energy R&D report lists the solar energy budgets of the most active countries:

'The main research effort for 1974 was made by the United States with $14 million followed by France ($3.9 million), Japan ($2.7 million), Germany (approximately $1 million in 1974 and $3 millions in 1975), Australia ($0.5 million) and Canada. The United

6

States research effort for solar energy represents 1.4 per cent of the total United States federal energy budget and constitutes a three-fold increase over FY 1973.'

In the International Energy Agency's allocation of different energy R&D sectors to various 'lead agencies', solar R&D went to Japan.

The spending of the US is really just getting under way. In all the NSF had a solar energy budget of $50 million for the fiscal year 1975. Congress is keen on solar power – a sizeable number of bills have been introduced to support solar energy work, ranging from $50 million for demonstration projects for solar heated and solar cooled buildings, to a $600 million across-the-board R&D effort over the next five years. Congress was very kind to ERDA when it asked for $70.3 million for solar R&D in FY 1976, and the House Science and Technology Committee boosted the amount to $143.7 million. In solar energy's dim years, between 1950 and 1970, the US Federal government invested an average $100000 a year on solar energy.

Chapter 9

'Earth power'

We have not reached the end of the list of 'alternative' energy technologies; but most of the less conventional energy systems can have an impact only at a local level. Fusion power and solar energy could dramatically alter the energy picture: no other new system offers quite the same potential as these. Fusion and solar resources are far larger than other alternatives; in fact, the Sun drives many of the 'second strings' in the energy repertoire. The advocates of wind power or tidal power, for example, are as vociferous as other energy researchers; but they cannot support their case with quite the same conviction. Few independent analyses show healthy prospects for the 'natural' energy sources. A realistic evaluation shows that while we can tap alternative energy supplies where they are relevant, they will never add up to much. At the same time they often demand significant social changes if they are to be widely employed. The decentralisation inherent in the development of some solar power ideas will have to be taken significantly further if wind power and other notions catch on. There are few signs that people are willing to accept such dramatic changes when they could meet their energy needs with less revolutionary ideas.

Geothermal energy

The OECD report *Energy prospects to 1985* says that 'the main foreseeable potential for non-conventional energy sources in the medium term derives from solar and geothermal energy, and energy from waste'. The WEC energy resources survey put geothermal energy into the 'limited but locally useful' category. In other words, geothermal energy may be relevant in those parts of the world where the heat of the Earth's core penetrates the cold crust and comes near to the surface. Thus volcanically active regions are potential geothermal areas, although it is hard to see how we could safely harness the energy of a volcano direct. Two varieties of geothermal energy are already tapped: dry steam, and hot water fields. A dry steam field is an underground reservoir of trapped water that is so hot that dry steam spews forth when the field is tapped. A hot water field yields either hot water or a mixture of hot water and steam. Dry steam is the most convenient geothermal resource; but various estimates suggest that in the US low temperature, hot water resources are twenty times as plentiful as dry steam fields.

Geothermal resources already generate electricity and supply hot water. Steam from a dry field can drive a steam turbine. In a hot water field, the steam and water have to be separated. The steam can then drive a turbine while the hot water meets some other energy need. The Icelandic capital, Reykjavik, is almost wholly heated by a district heating scheme fed from geothermal wells. But geothermal hot water cannot be economically transmitted over any distance, and thus it is only useful where there is a local demand.

Water and steam are not the only purveyors of geothermal energy. Plenty of energy is tied up in hot dry rocks. As yet this particular geothermal resource has not been exploited. This must wait until the requisite technology is developed. The way to the large energy resource of the hot rocks is to drill a hole into the rock, shatter it so that water can be pumped down into the well and brought into contact with a large surface area, and finally to pass through a steam turbine the steam generated when water and rocks meet.

A fourth variety of geothermal energy is the 'geopressure zone'. Here too water is the energy carrier, but its energy is not heat energy. Water trapped underground may be at an extremely high pressure. If an underground pressure zone is penetrated, a huge mechanical force is released. As well as its hydrostatic pressure, the underground water zone may be hot, and it may contain dissolved natural gas. Joseph Barnea, a United Nations expert on geothermal energy, has said that: 'We know today that there are vast areas extending over about 200 000 square miles of geo-pressure zones in Louisiana and Texas and some other parts of the US. Geopressure zones are also known to exist in many countries in Europe including offshore areas, practically every-where where petroleum drilling has been undertaken.' He says that geopressure zones could be widely utilised if the appropriate technology is developed. The novelty of this variety of geothermal energy, and the fact that it is the oil companies who have the greatest experience with geopressure zones, means that little is known of their extent or possible contribution to world energy needs.

The exact depth at which geothermal energy resources are found varies, but the usual depth-spread is between 500 and 2000 metres. (It can be as shallow as 50 metres, or as deep as 3000 metres.) The viability of geothermal energy depends upon the economics of drilling to such depths. The deeper you go, the higher the cost, hence the economics of geothermal energy exploitation must depend on the depth at which the energy is found. Clearly the best people to do the drilling for geothermal resources are the oil companies, with their extensive knowledge of the drilling business. They must, however, adapt their technology to the needs of geothermal energy exploiters.

There are several obstacles standing in the way of geothermal energy utilisation. Geothermal resources may be cleaner than some energy sources, but they are far from pollution free. The steam in a geothermal field usually contains some hydrogen sulphide and other pollutants. Not only does the hydrogen sulphide corrode the materials in a geothermal energy plant, it also smells bad. It can cost something like a million dollars to suppress the odour of a geothermal energy plant. Another environmental hazard is the salt that may be dissolved in the water of a geothermal field. If there is a lot of salt in geothermal hot water the waste water cannot be dumped into a river unless it is very near the sea. To prevent the possibility of undesirable environmental damage, it may be necessary to develop technology that allows the waste water to be re-injected into the underground reservoir.

The problems of geothermal energy have not prevented the development of attractive resources. Electricity was first generated from geothermal energy at Lardarello in Italy in 1904. Since that experimental venture the generating capacity of the field has been increased to around 400 MW e. Japan's use of geothermal energy amounts to more than 30 MW e; and in the Californian Geysers

▶ Electricity was first produced from Italy's Lardarello geothermal energy field in 1904. Lardarello is one of several geothermally active areas in Italy – the steam, which can be tapped by drilling down 700 to 1600 metres, is at 200 °C and a pressure of 5 atmospheres. The steam contains gases and chemical impurities which are a valuable chemical feedstock. Each Lardarello steam well lasts about 20 years before it cools down, although some wells have produced steam for 30 years. The Lardarello power generating plant was destroyed at the end of the Second World War, but subsequent repair and development has raised the output of the region to 400 MWe.

the generating capacity is more than 500 MWe. These three countries have the good fortune to have dry steam fields. Other countries have to exploit hot water reservoirs to generate electricity:
these include New Zealand (290 MWe) and Mexico (75 MWe).
Thus the world's installed geothermal generating capacity adds
up to the equivalent of a couple of medium-size nuclear reactors.

Geothermal resources can provide hot water as well as electricity. District heating is one way to use geothermal heat; in New
Zealand a pulp and paper mill generates electricity and operates
manufacturing processes with geothermal heat. Desalination may
also be possible with geothermal energy. Underground hot water
can, with the right technology, be cleaned of its impurities to
yield fresh water. As yet there is no commercial desalination plant
in operation at a geothermal field, but the United Nations Geothermal Project is supporting trials with a desalination pilot plant
at the El Tation geothermal field in Chile, where electricity generation is also a part of the project. Geothermal waters could provide
useful minerals as well as fresh water and heat. These minerals
have been dissolved from the underground rocks which the waters
have come into contact with. Geothermal water may not be hot
enough for many industrial processes, but agriculture can put
warm water to good use. Geothermal heat has been employed
to warm greenhouses, and to heat animal buildings and fish
ponds.

What are the technological problems facing the would-be user
of geothermal energy? The OECD study of energy prospects to
1985 said: 'Exploitation of geothermal sources requires some
R & D, particularly in overcoming corrosion and disposal problems
connected with hot brine. Injection of pressurised water to depths
where it can be heated by hot rocks is a technique worth
evaluating in view of the large hot dry rock reserves available.' An
important part of any geothermal energy programme must be
a resource identification effort. Because geothermal resources
have yet to be exploited to any extent there has been little
incentive to locate promising reservoirs. The OECD study points
out that: 'A better assessment of the potential of geothermal
deposits does in fact require a major research effort aimed at
improving detection and exploration techniques, as none of the
present methods, including geophysical, geochemical techniques,

and electrical measurements appear to yield comprehensive results.' Limited prospecting may be possible with aerial photography, using infrared and microwave measurements to look for thermal zones near the surface. The Skylab space shot looked for geothermal sites during its experimental programme.

The US Geothermal Energy Act set a target of 20000 MWe of generating capacity by 1985. This was also put forward in WASH-1281, which added that the target for the year 2000 should be 200000 MWe. A report prepared by the US Geological Survey recently classified about 1.83 million acres (7500 square kilometres) as being 'known geothermal resource areas'. The same study said that an additional 99 million acres (400000 sq.km) are considered to have 'prospective value' for geothermal steam.

A study by Britain's Central Electricity Generating Board said: 'Our estimates of plant costs, more than five times that of fossil fired plant and more than three times that of nuclear plant, appear to be daunting, but it must be remembered that geothermal power has zero fuel costs.' The study goes on to say that the only way to evaluate the economic viability of a 'zero fuel cost' energy system is to 'estimate the present day worth of the power produced over the lifetime of the plant it would replace'. Unfortunately, the abundance of unknowns makes it difficult to arrive at a meaningful assessment of the prospects for geothermal energy in Britain, 'but it seems that geothermal power would not be economic at present but it could become so with improvements in performance and a significant increase in nuclear and fossil fuel prices'.

A UN estimate of the cost of geothermal electricity, published at the end of 1974, said that for plants now in operation its cost is 'less than one cent per kWh' of electricity produced. Nuclear costs, thanks to operational difficulties, are 'significantly in excess of one cent per kWh at most nuclear power stations'. And the cost of generating electricity in oil-fired power stations is about two cents per kWh. As for low-grade thermal heat from geothermal sources, this is 'of the order of one-tenth, or less, of the cost of heat using fossil fuels at current prices'. Joseph Barnea has said that in California a comparison between different electricity generating systems revealed that 'alternative sources of generating power using fossil or nuclear fuel would be approximately 1.5 to 2 times the cost of geothermal electricity generation'.

This estimate is for a dry steam field (the Californian Geysers). It would cost more to develop a dry rock field, because of its complexity. The cost of drilling the geothermal well might be the same; but the rock has to be fractured (this can be done by nuclear explosives or by pumping water into the reservoir under pressure, hydrofracturing) and water has to be found and pumped down into the well, making the system more complicated than a wet well. The costs of these associated activities will determine the viability of this particular type of geothermal energy.

One experiment to develop hot rock technology was carried out in 1974 with National Science Foundation funds in the US. As a part of a $2\frac{1}{2}$ million project a hole was drilled in Montana to look for hot, dry rock. The intention was to drill down nearly 2000 m, or until they encountered rocks at 420 °C. This experiment went wrong when the drillers struck massive quantities of water.

Dry wells would have at least one advantage over wet geothermal resources. The steam they generate will probably be less damaging to turbines than steam from a wet well, with all its corrosive impurities. The corrosion problem will be somewhat alleviated thanks to technology developed by the oil industry which has to cope with 'sour wells' where the oil contains a lot of sulphur.

Technical problems are not all that the would-be geothermal electricity utility has to face. Electricity generation based on the Earth's heat brings new difficulties that are normally associated with other energy systems. Oil companies may be attuned to the high risk of failure when they drill a new well, but it might be too much for an electricity company to drill a geothermal well only to find that there is nothing there. Risks like this are new to the utilities. To develop a geothermal generating capacity of 200 to 400 MWe, the smallest unit that would interest the utilities, a utility would have to drill something like a dozen wells in a successful geothermal area to obtain enough energy. And with the cost of a well working out at something like $\frac{1}{2}$ million, the electricity utilities may be reluctant to look for 'free' energy.

As with other new systems, there are various legal obstacles to be cleared aside before geothermal energy can be harnessed in many countries. The US has its Geothermal Energy Act; but other nations are not likely to deal with legal aspects of geothermal

▶ 'For God's sake shovel faster Bert.
They're gaining on us.'

energy until they are forced to. Few independent companies would embark upon an extensive development programme without some assurance that they would not face enormous legal obstacles at the end of their quest.

Garbage power

Energy from waste must stand alongside solar energy and geothermal energy in any assessment of unconventional energy sources. Scientists and engineers may prefer to work on more glamorous projects than the development of garbage incinerators, but 'waste power' meets several needs. Not only does it supply energy, it also reduces the growing problem of what to do with the never ending piles of rubbish we expect to miraculously disappear once our dustbins have been emptied. 'Urban ore', as garbage has been called, can be burned to yield energy or it can be processed to produce fluid fuels. And the technology needed to do these jobs is either already available, or the subject of on-going R & D programmes.

Britain disposes of 20 million tonnes or so of garbage each year. This could provide the equivalent of 6 million tonnes of coal. (In 1972 the UK electricity supply industry consumed the equivalent of 110 million tonnes of coal.) There are various estimates of the energy that could be squeezed out of the US's annual output of 200 million tonnes of municipal garbage. Alan Ferguson, an industrial economist with the Stanford Research Institute, estimated that in 1974 'on a national basis, the refuse would appear to be able to supply about 10 per cent of the utilities' fuel requirements'. For the State of California garbage might provide nearer 15 per cent of the fuel for the electricity supply industry. The report *Canada's energy opportunities* says that 'Alberta's Agriculture Department has calculated that fermentation of the province's annual production of cattle manure alone could supply 6 per cent of the province's present gas consumption.' An EEC study says that: 'After biological processing, the solid waste from a city of two million inhabitants could supply sufficient material to feed a 1000 MW electrical power station.'

Waste already provides energy in some cities. Frankfurt's waste yields around 7 per cent of the city's electricity. Amsterdam obtains

6 per cent of its electrical output by burning waste. Germany has over 20 refuse-burning installations; and in Vienna industrial steam is made by burning waste. In Toronto, a 300-unit apartment block is being designed so that all of its hot water needs can be met by burning garbage.

As this list shows, most of the experience in garbage energy recovery has been with incineration systems. In some incinerators garbage is merely an additive to fossil fuels. Alternatively, wastes can be burned on their own, often in a modified coal incinerator. This form of energy recovery has become more attractive as the cost of garbage disposal has risen. And with the added incentive of higher fuel prices the economics of energy recovery from waste have become even more appealing. There are, however, ways of turning urban ore to good use which are less costly than just burning it. These alternatives require some R&D before they can compete with simple incineration.

There are several techniques that can be used to extract energy from waste material: conventional incineration; incineration in power station boilers; fluidised-bed combustion (see chapter 4); pyrolysis; hydrogenation; and anaerobic digestion. Of these possibilities incineration is already well developed, but it is not necessarily the most satisfactory because it can be costly and because there are more efficient ways of extracting energy from wastes. Of the undeveloped technologies, pyrolysis is being pursued with greatest vigour. Pyrolysis is the process of 'cooking' the material at a high temperature in an oxygen-deficient atmosphere. (It is a process that is also being looked into as a way of converting coal and plant material into other fuels.) Britain's Warren Spring Laboratory – an outpost of the Department of the Environment – is one research establishment working on pyrolysis, which is really a waste conversion process rather than a direct energy conversion system. Biological processes, such as anaerobic digestion, have a similar goal. Here the aim is to 'feed' microbes on garbage; methane is generated in the process. Hydrogenation is similar to pyrolysis. Otherwise known as chemical reduction, this technique calls for the organic waste to be 'cooked' in the presence of water, carbon monoxide, and a catalyst.

A typical pyrolysis system is the Landgard process developed by Monsanto Enviro-Chem Systems Inc. Waste is heated in a kiln to yield a low-Btu gas (100 Btu/cu.ft). Monsanto has tried out the process in a 35 ton/day demonstration plant in St Louis County, Missouri. Not only does the Landgard process reduce the volume of waste (by about 94 per cent once the ferrous metals have been removed), it also produces a low quality gas. This gas cannot be economically transmitted over any distance; it has to be burned near the waste processing facility. In Baltimore a 1000 ton/day plant has been built near a power station operated by the Baltimore Gas & Electric Company. The waste facility cost $15 million. Monsanto estimates that the cost of a Landgard system would be about two-thirds the cost of a conventional incinerator, a figure that is supported by other estimates.

Pyrolysis plants can make oil or gas. The Garrett Research and Development Company, a subsidiary of Occidental Petroleum, has developed a waste pyrolysis process that yields something the company calls 'Garboil'. Garrett's approach resembles Monsanto's, but the feedstock has to be more carefully sorted before processing. (The waste that goes into making Garboil has to be 95 per cent organic material.) Garrett claims that each ton of garbage yields a barrel of oil. But the oil has only three quarters of the energy content of a barrel of fuel oil.

Fuel production has distinct advantages over simple incineration. Not only does it cost less to build a pyrolysis plant, but oil can be stored until it is needed. With an incinerator the only way to store energy is to accumulate piles of rubbish. And incinerators have trouble when they burn materials such as polyvinyl chloride, which gives off corrosive acid vapours. On the positive side, waste generally contains less sulphur than fossil fuels, which means that it presents less of an air-pollution problem.

Pyrolysis is not yet a 'proven' technology. However, the current round of demonstration plants should confirm the feasibility of the technique. They should also confirm the cost estimates based on smaller units. Techniques such as hydrogenation are not so well understood. One guideline to the competitive status of a synthetic oil plant is the cost of the product in comparison with the genuine article. The OECD report on energy R&D says that: 'Oil produced from urban waste [by pyrolysis] has been estimated to cost $8.4 per barrel in a 1000 ton per day plant, whilst the estimated cost per barrel rises to $16.6 for the conversion of

purposely grown organic matter.' The same report puts the cost
of oil produced by the hydrogenation of wastes at around \$4 to \$5
per barrel, 'taking into account the cost of incineration waste
disposal methods averaging \$10 per ton'. The report says that
laboratory experiments have produced yields equivalent to $1\frac{1}{4}$
barrels of oil from a ton of dry waste. It adds that: 'Economic
feasibility of this process is however to be more accurately evaluated because many technical problems are still to be solved and
because of the difficulty of assessing at present the operating cost
of a commercial size plant.'

Organic waste can also be turned into a useful fuel by bacteriological decomposition (anaerobic digestion). Excreta and veget-
able wastes will yield 'bio-gas' – a mixture of methane and carbon
dioxide – if the material is put into a container that keeps out the
air but releases the gas given off by decomposition. Sewage works
have used 'digesters' to generate their energy needs for some
time; but bio-gas has mostly been made on a small scale by enthusiasts. One of these pioneers, L. John Fry, experimented with
bio-gas units for more than 20 years. Fry started his experiments
when he operated a pig farm in South Africa. He didn't like to
see two tons of pig manure a day going to waste, especially as he
had to find a way of getting rid of the manure. Fry built a succes-
sion of larger and larger waste digesters. In the process he came
up against, and solved, many of the problems of bio-gas manufacture. For example, a solid scum forms on the surface of the waste
material in a digester. This scum accumulates until eventually it
halts gas production. Fry developed a simple scum removal
system.

Thanks to a long series of experiments by people like John Fry
bio-gas production is now a fairly widespread 'alternative
technology'. It has yet to make the transition to an industrial scale,
but there is some sign of growing interest in the agricultual community. Farmers like the idea of cutting their energy bills; and
they also like the idea of solving their waste disposal problems.
(Farmers are among the greatest polluters of inland waterways
because their manure collection tanks have a habit of spilling
their contents into nearby rivers.) There is an added bonus that
comes with bio-gas production: the solid material that is left over
after bacteriological decomposition can be spread on to the land

as a fertiliser. Bio-gas production also appeals to the small communities of enthusiasts who want to cut themselves off from the existing energy system. One British company sells a series of bio-gas makers. Small units like these will not put the natural gas industry out of business, but they can supply the energy needs of some small communities.

Power from the elements

The list of energy 'also rans' contains a group of energy sources that are 'natural' and continuously refuelled by the Sun. Wind power, wave power, and tidal power are on the list along with the sole 'renewable' energy source that is already employed on a worldwide basis, hydroelectric power. In all cases the weather does the work that can be turned into useful energy. The OECD energy R & D report says of these energy systems: 'Although it is true that all these sources have a high energy potential, it is too early to determine what their contribution could be and whether they can be exploited economically. Some of them appear nevertheless to be fruitful areas of research.' Dixy Lee Ray's WASH-1281 ignores the sea's potential contribution – it does not mention wave and tidal power – and advocates a relatively minor programme for wind energy, which it lumps in with the solar energy programme. The report advocates an expenditure of $31.7 million over five years. This is a little more than it says the US should spend on ocean thermal energy, and a little less than for photovoltaic conversion.

As with geothermal energy, the World Energy Conference is very vague about the possible level of the natural energy resources. It says that the 'total power of the wind in the entire atmosphere is estimated to be 2600×10^{12} gigawatt-hours/year of which about one quarter is over land'. It then says that the estimates of usable wind energy – which range from 1 to 20 million megawatts – 'may be extremely conservative'. Tidal power, the WEC study says, could supply up to 13–20 GW, with an average annual output of around 175 000 GWh of electricity. Wave power is not mentioned at all in the WEC study; indeed, it was not taken at all seriously until relatively recently, when the British government started putting money into the subject.

Apart from the occasional tidal power project, the only renewable source that is operated in large units is hydroelectric
power. The development of further hydroelectric power capacity
depends less on new technology than on the availability of good
sites that are not too far away from a market for electricity. The
WEC study says that 'ultimate hydroelectric output, if all sites
were fully developed, would be about 10 million gigawatt-hours/
year of which about 46 per cent would be available 95 per cent of
the time'. The study puts current world output at 1.3 million
GWh per year, 'about 13 per cent of ultimate potential output'.

Hydroelectric power requires little R & D to help it on its way.
Potential sites have to be found, but once that is done civil engineering takes over. (Research that would improve our knowledge of geological processes might reduce the possibility of a
catastrophic collapse of a hydroelectric dam, but this is not a pressing problem that warrants any special effort.) The developed
nations, where the need for new energy sources is greatest, are
probably less capable of increasing their hydroelectric capacity
than the developing countries whose potentially significant hydroelectric sites have yet to be exploited.

Tapping the tides

Tidal power is a variant of hydroelectric power that could tap
one of the world's natural energy systems. The pull of the Moon
and the Sun shifts huge quantities of water around the Earth's
oceans, resulting in some hefty tidal rises and falls where local
conditions are right. If this tidal flow is made to fill and empty
large reservoirs, it can create a difference in levels between the
water in a reservoir and the sea. This level difference provides a
hydrostatic head that can drive a water turbine, much as a hydroelectric project derives its power.

Few places have the right conditions for tidal power generation.
That is, few locations have a large enough difference between high
and low tides, and the necessary geographical configuration that
allows a reasonably large water reservoir to be enclosed by as
short a dam as possible. The large reservoir is important because
the power that can be generated depends upon the amount of water
contained by the dam. A short dam is necessary to keep down

the cost of construction. And without a large enough tidal range you can forget the whole idea. The Bay of Fundy, in Canada, is often talked of as a promising site: it has tides that average 11 metres. The report on Canada's energy opportunities said that 'in theory, systematic development of potential sites could provide the Maritime Power Pool with most of its electrical energy needs for this century'. The Severn Estuary is Britain's most promising site for tidal power. As yet the only tidal power project that has produced any significant experience is the French project at La Rance, in the north-west part of the country, where peak tides of up to 13.5 m occur.

The French project is of the simplest variety: a large dam, about 750 m long, encloses a sizeable reservoir which fills as the tide rises and empties again as the tide falls; sluice gates close off the flow through the dam to produce the water head to drive the station's turbines. Work started at the La Rance site in the middle of 1960, and full scale construction work started in January 1961. The estuary was finally closed off from the sea on 20 July 1963. The first generating unit was started on 19 August 1966, and the last unit on 30 November 1967. The final construction cost was 570 million francs, which when corrected for inflation represents a cost overrun of 14 per cent. The power station produces a little more than 500 GWh a year, and the mean capacity throughout the year is 65 MW.

La Rance exhibits the characteristics of any single-basin tidal power project. The availability of power is dictated by the tides, which rarely match the demand for electricity. This, and the high cost of tidal projects, weighs heavily against them. The French project has proved that there are few technological difficulties. The 24 turbines are used both for generation and for pumping water into the reservoir when there is excess power available elsewhere in the network. The French have made great progress in controlling corrosion of concrete, metals, and the coatings that are applied to them. The technology is proven; and yet the French have not gone ahead with any of the proposals for extending the scheme.

We cannot hope to change the tides to match supply from a tidal scheme to demand, but we can use spare generating capacity to store energy by pumping water up to a higher level. Pumped

► The tidal power station at La Rance in France is proof that this energy technology works. The dam is some 750 metres long, and the station started generating electricity in 1966.

storage is becoming common practice and will spread as nuclear power becomes more common. (You cannot switch off a nuclear power station in the same way that you can a fossil-fuelled station, so it is easier to leave the station running and to use its output to pump water that can subsequently be released in a hydroelectric scheme.)

However, at La Rance they use another approach. Spare generating capacity elsewhere in the local supply system is used to run the tidal project's turbines as pumps. In this way they have boosted the mean output at La Rance from 45 to 65 MW. Sometimes the pumping power may be as high as 150 MW. And the tides can multiply the input energy; at La Rance 1 kWh of electrical input yields 2.8 kWh of output. But this technique does not get around the tidal mismatch. A single reservoir system can never overcome this drawback. Demand and supply can be matched by adopting more sophisticated reservoir configurations. An electricity utility cannot rely on having a simple barrage tidal power unit such as La Rance available when electricity demand is at its peak – the tides will rarely coincide with the peak demand for electricity. So with this sort of tidal barrage, spare generating capacity has to be available for times when the tides are wrong. In other words, tidal power can save energy but it cannot save capital investment in more conventional generating equipment unless the power can be supplied when it is wanted.

Several possible reservoir configurations have been proposed for the Severn Estuary. Their aim is to put the output from the project under the control of the operator rather than the tides. The Severn has been described as one of the best tidal power sites in the world. T. L. Shaw, of Bristol University, says that 'two basins could be constructed in the estuary, each filled and drained when the tide is high and low respectively; energy would be generated by flow between the basins'. This idea was put forward 30 years ago, during one of the regular reappraisals of the Severn project. Shaw's two-basin system has a smaller low-level basin downstream of the large basin. The large, high-level basin would run across the width of the Severn Estuary, while the low-level basin would be circular, and would have a common wall with the large basin. The actual shape and size of the two reservoirs would depend upon the required power output, and the geology of the estuary. Nighttime pumping could lower the level in the small reservoir, while increasing that in the larger one. Shaw estimates that such a scheme could produce 4000 MW at a constant rate over the 12-hour period each day when the power is needed. Such a project could provide 6 per cent of the UK's day-time electricity consumption in the mid-1980s.

The CEGB has looked into various tidal power schemes. In 1974 it concluded that the Bristol Channel scheme (which is the Severn Estuary project under another name) could supply 12 per cent of the country's present electricity demand 'at a cost which might now be economic'. But before we could embark upon an ambitious project such as this, we would have to set in hand a more detailed technical study as well as a thorough environmental evaluation of the scheme. We do not know what a large disruptive project might do to the ecology of the Severn Estuary and the Bristol Channel. The zero fuel costs of a tidal power project are attractive, if they can be achieved at a reasonable capital outlay. But, as the CEGB says, 'few suitable sites exist either in the UK or the world, so the total contribution to future energy demands would not be large'. Recent events have done little to change the technological status of tidal power projects, but they have sent those countries with potentially good sites back to reassess the economic prospects of the various schemes.

The crest of a wave

Wave power is one of the few natural energy systems that is almost solely a British interest. Small machines that can tap the power of the waves by bobbing up and down are not a new idea. A huge structure that can equal the output of a power station is more original. This completely untried idea gained official support while other, less extreme (but still way out) possibilities got short shrift in the report of the UK government's Central Policy Review Staff on energy conservation (see chapter 3). This report said that wave power 'has some favourable features in the United Kingdom context. The usable coastline amounts to some 900 miles and the energy theoretically available from its exploitation on a reasonable economic basis would sustain a capacity of up to 30,000 MW' (about half the CEGB's installed capacity at the time). The UK government commissioned a study of wave power from the National Engineering Laboratory. Armed with the preliminary findings from this study the government awarded a £65 000 research grant to Dr Stephen Salter of Edinburgh University. Even the CEGB acknowledged that 'the practical difficulties are daunting and the costs, at present speculative, appear uncompetitive. A research programme is nevertheless justified to reduce the uncertainties.'

Salter's innovation was in the way that he proposed to extract the power from the waves. Waves bobbing up and down may look 'energetic', but Salter's first move was to throw out the idea of a system that relies on something rising and falling with the waves. He wanted to develop a system that would harness the to and fro motion of the waves. He designed a vane that could rotate with the motion of the waves. Salter did some early work on vane designs, and as a result of the government research grant he has moved on to model tests in a tank that simulates the way that waves move. Even at this level of funding Salter has to work with models that are only a hundredth the size of the real thing. Anything nearer life size would require significantly more money.

There are other proposals for wave-power systems. Indeed, there is even a commercial company working on wave power. Wavepower Ltd has yet another approach to wave power. It wants to build strings of floats that are hinged together. As a wave passes the line of floats it would cause them to rotate relative to each other. If the 'hinges' between floats can operate air pumps, for example, energy can be extracted from the waves.

Wave power could be turned into usable energy in a variety of ways. Salter suggests that high pressure water pulses could take power from the rotating vanes. The energy could be sent ashore by generating electricity on board the wave-power system's floats and relaying it by cable, or by turning water into hydrogen by electrolysis of sea water. The CEGB put forward another alternative: it suggested that the power could be used to separate uranium from sea water (see chapter 6). 'Cheap and virtually limitless supplies of uranium produced in this way could have a profound effect upon the future of nuclear power; for instance fast-breeder reactors would no longer be essential for nuclear fuel economy.' (The CEGB appears to have an aversion to breeder reactors; it pounces on any development that could make them unnecessary.)

► Research on wave energy is at an early stage. Scientists at the Central Electricity Generating Board's Marchwood Engineering Laboratories, near Southampton, England, have tested the ideas put forward by Dr Stephen Salter of Edinburgh University. They found that Salter's proposal for a wave tapping device could be very efficient, taking more than 90 per cent of the energy out of a wave.

A wave-power 'station' would be a huge structure, but it could be technologically quite simple, far more so than a nuclear power station. A wave-power station would probably have to be between $\frac{1}{2}$ and 1 kilometre long if it is to be stable in heavy seas. And it would have to be anchored to the sea bed; or it could be towed way out to sea so that it could gradually move along with the waves, tapping power as it goes. The cost of a wave-power system might be high. The CPRS report estimated the capital cost at around £420/kW. The CEGB estimated that 'the cost band from £400 to £800/kW may well cover the eventual figures'. This is well above the cost of a nuclear power station. But it might be possible to build a wave-power unit in 5 or 6 years, while nuclear power stations can take 6 to 10 years with a large tidal project taking nearer to 12 years.

Neither the lead time nor the free fuel were particularly significant factors when the UK government decided to back this technology. Both the Department of Energy and the CEGB see wave power as something of an insurance policy, taken out just in case there is a disastrous nuclear accident that makes it impossible to continue building nuclear power stations. The CEGB is a little more delicate: 'The need for an insurance policy to guard against possible adverse circumstances in the future justify a small programme of research in this area.' As a result the CEGB set up its own small wave-power research programme. This insurance policy appears more attractive to Britain than some of the other 'natural' energy systems because, unlike solar power for example, Britain receives more wave energy off its shores in winter, just when energy demand is highest. Salter's machines could, if early experience transfers to larger machines, tap the power of the waves with a remarkably high efficiency. The aim of the work is to develop systems with an ultimate efficiency of between 60 and 70 per cent in taking wave energy from the sea to the consumer. This far outweighs the efficiency that other natural energy systems look like achieving.

Blowing in the wind

Wave power is really no more than a way of tapping the energy of the winds. Wind power has never ceased to arouse the interest

of the lone inventor and the would-be energy isolationist. Some of
these fervent advocates of wind power have managed to persuade
people to spend money on some sizeable wind-power units. As
with solar power, most of these early experiments never got beyond
the R&D stage. The situation is changing and detailed R&D
programmes are being undertaken on a wider scale, most notably
in the US where NASA and the NSF responded to the energy
crisis by including a significant wind-power programme in their
joint solar energy effort. (It is, after all, the Sun that creates the
wind.)

Most wind-power advocates believe that viable systems can be
developed with strictly conventional technology. Propellers will
rotate in the wind; so all you have to do is build a propeller of the
right shape and size to provide the amount of energy required.
However, one of wind power's big problems is its intermittent
nature. The cost of wind energy is, according to the Canadian
report on the country's energy prospects, 'two or three times the
cost of conventional energy in Canada's populated areas'. Much
of this cost is taken up by the energy storage system that supplies
power during calm periods. The Canadian report adds that wind
power could be competitive in remote parts of the country 'particularly if the cost of purchasing and transporting conventional
fuels increases and as the technology for storing energy improves'.

Once again the US has done most to evaluate the potential of
wind power. The NSF/NASA report on solar energy estimated
a total possible wind-power production of 1536 MWh per annum
by the year 2000, and called for an R&D programme costing
$610 million over 10 years. The US's wind-power R&D budget
has risen dramatically from $200000 in the 1973 fiscal year to an
estimated $12 million in FY 1976. Estimates of the amount of
energy tied up in the winds are not really relevant to the viability
of wind power. There is enough wind energy available to make
a significant contribution to our needs, but this will be made only
if economical power stations can be developed. Wind energy has
one significant advantage over the Sun in Britain – like wave
power it is at its most intense in the winter when the country needs
the energy.

The energy in the wind depends, quite obviously, on the speed
of the wind; but it is not a simple linear dependence. As the wind

speed rises the energy content goes up as the cube of the wind velocity. Not all of this energy can be recovered by a rotating windmill. A wind-driven device can recover a theoretical maximum of only 60 per cent of the energy. Mechanical losses will reduce the energy recovery to a maximum of about 40 per cent, probably less. We must also remember that maximum winds are the exception rather than the rule. Professor Karl Bergey, of the University of Oklahoma, pointed out some of the realities of wind energy in a paper prepared for the energy sub-committee of the United States House of Representatives Committee on Science and Astronautics. Bergey told the committee that it does not make sense to build a windmill for maximum wind speeds; 'it is more cost effective to use a small generator which maintains a constant output at all speeds above its design wind velocity'. We can extract 90 per cent of the realisable wind energy at a particular site (which is, as we have seen, a fraction of the total wind energy) by building a wind rotor that is only half as big as a rotor large enough to realise the full potential of the site.

A rotor will turn only when the wind is blowing. According to Bergey: 'a typical wind generator operates at an overall load factor between 20 and 30 per cent, only one-third that of a typical fossil-fuelled plant. Thus, for a given annual power output the wind generator will need approximately three times the installed capacity of a conventional steam plant.' Bergey's experience is with the winds of Oklahoma, but his conclusions are valid wherever the wind blows.

Experiments with wind-driven generators go back a long way. In the first decade of the twentieth century Denmark made great use of wind power, using 33 000 generators each producing 35 kWe. The largest wind rotor was built in Vermont in the US and operated between 1941 and 1945. This was the $1\frac{1}{4}$ MW generator with a 175 ft diameter, twin-bladed propeller, built at Grandpa's Knob and designed by Palmer Cosslett Putnam. In 1942 this generator produced 179 MWh of electricity. And in 1945 the unit operated for a month as a generating station in the supply system of the Central Vermont Public Services Corporation.

The Vermont project came to grief in March 1945 when one of the eight-ton blades snapped, forcing the S. Morgan Smith Company, which had spent $1¼ million on the project, to abandon its work on windmills. Metal fatigue was found to be the culprit when the broken blade was tested to find out why it failed. Had there been any chance of turning wind power into a commercially viable system, it would have been easy enough to overcome the fatigue problem. In the 1940s and 1950s the energy situation was not propitious for the development of wind power or other unconventional ideas. To begin with, oil was catching on as prices fell; and scientists were more interested in the challenge of nuclear power, which was far more glamorous than wind power.

Now that the world looks longingly at almost any source of energy, wind power is back in favour. The programme initiated by the US National Science Foundation covers a variety of ideas and projects. The National Aeronautics and Space Administration plays a leading role in much of this work. As the home of knowledge on aerospace technology, NASA clearly has a lot to contribute on topics such as propeller design. However, there were some objections when NASA was brought into the wind-power programme. Karl Bergey warned that if this high-technology industry takes the reigns 'there is too great a danger that the results would be overly complex and overpriced'. It is not technological sophistication that is needed to make wind power attractive; far from it, inexpensive and simple technology stands a far better chance of success.

The US has just begun to spend money on large wind-power projects, so it is too early to know how the technology will develop. It might prove feasible, for example, to build a lot of small windmills; or it may prove preferable to build a few large devices. One of NASA's earlier moves was to initiate a project whose aim was to develop a small and simple windmill based on a 50-year-old idea. The Langley Research Center announced early in 1974 that it was working on a vertical axis windmill that resembles half an egg beater. Patented in 1927 by the French inventor George Darrieus, the windmill has two blades, curved like hunting bows, attached at top and bottom to a vertical shaft. As the wind blows the rotor turns about the vertical axis. (A conventional windmill has a horizontal axis.) The rotor is connected to a generator by a simple gear mechanism. The blades, which are aerofoil shaped, rotate even in very light winds. And no matter what direction the wind blows the rotor faces the wind. Canada's National Research Council supported earlier work on the Darrieus rotor by Raj Rangi and Peter South, whose work was taken up by HPL Engineering of Ottawa, as well as the Langley Center. The Canadian company built eight-foot windmills, while NASA's device was twice this size. NASA estimated the cost of building and installing a windmill for a house at between $500 and $1000. HPL put the cost of its device between $1000 and $1500, including the generator and battery storage for times when the wind is not strong enough to generate sufficient power.

Another small windmill, of a more conventional design, was built by the Grumman Aerospace Corporation. Grumman's 'Sailwing' was based on a design of a semi-rigid aerofoil developed at Princeton University. Sailwing's 25-ft diameter three-bladed rotor was put atop an aluminium mast. Thomas Sweeney, who invented the Sailwing's simple but elegant lightweight aerofoil with its covered construction rather than solid blade, estimated the possible cost of a complete system at $4000 or so in 1974.

The most advanced of the larger windmills is the Lewis Research Center's project on the design, construction, and operation of an experimental 100 kW generator at a cost of around $1 million. This windmill, located at the NASA–Lewis Plum Brook test area, Sandusky, Ohio, has 125-ft diameter blades and is mounted on top of a 125-ft tower. The wind turbine generator system consists of a rotor, transmission, alternator, and tower. The rotor blades are on the downwind side of the tower. The alternator and transmission equipment are in an enclosure on the top of the tower. Sensors and servo-driven controls rotate this upper assembly as the wind changes direction. The generator is designed to achieve an output of 100 kW when the wind speed is 18 m.p.h. Wind conditions at Plum Brook are such that the windmill should generate 180 MWh per year. At an 18 m.p.h. wind speed the rotor spins at 40 rev./min.; the transmission steps the shaft speed up to 1800 rev./min. When the wind speed rises above the design level, the rotor blades spill wind by increasing their pitch, maintaining the output at 100 kW. When the wind rises above 60 m.p.h. the blades are feathered to avoid any possible damage.

This project, the largest windmill built in the US for 30 years, was followed up by two $\frac{1}{2}$ million contracts to industrial firms

▶ Thirty years after Palmer Cosslett Putnam built an experimental wind-powered electricity generator in Vermont, the US started up 'the world's largest wind turbine' which has less than a tenth the output of Putnam's machine. The new wind turbine was built by NASA at the Plum Brook Station in Sandusky, Ohio. The 100 kW turbine is mounted on a 125-ft high tower, and has two 62½-ft long aluminium blades. The machine starts generating power in an 8 m.p.h. wind. The blades can rotate at a maximum speed of 40 rev./min. When the wind is faster than 40 m.p.h. the blades are automatically feathered and the turbines shut down to prevent any damage to the machine.

for preliminary design work on very large windmills for electricity generation. Under these contracts the General Electric Company of the US, and Kaman Aerospace Corporation are examining windmills capable of generating between 100 kW and 3 MW. A 3 MW windmill would be far larger than anything previously built: it would have rotor vanes approximately 200 ft in diameter and an electrical output sufficient to meet the needs of 100 to 200 American homes.

Sweden was also quick off the mark with wind power. In the middle of 1974 the Swedish State Power Board started an investigation into the feasibility of heating homes with wind power. The Swedish calculations showed that with a tower 60 metres high, and a propeller 57 metres in diameter, a windmill could produce 2000 kW at a wind speed of 15 metres per second ($33\frac{1}{2}$ m.p.h.). This would generate an average annual output of 4100 MWh. The State Power Board pointed out that 1500 stations like this would be needed to replace a single 1000 MW nuclear power station. Each wind-power station might cost $$1\frac{3}{4}$ million to build; but to be competitive the cost should be a quarter this estimate.

Windmills have been designed and built before, but technology has come a long way since those earlier experiments. Hence the need for a programme of R&D aimed at finding modern solutions to the problems of windmill design. These problems include those of blade design and construction, and control systems. A windmill must have light blades, but they must be strong enough to survive the strong forces that develop at their fast-moving wing tips. And blades must be able to withstand any vibrations that might set up in them. Metal blades are the best understood, but composite materials could ultimately prove to be the most economical for blade construction. Windmill controls include a system for changing the pitch of the blades to keep the rotor turning at a constant speed as the wind velocity alters. It is not too difficult to dream up complex electronic systems that can carry out these control functions, but what are needed are simple, foolproof, reliable, and inexpensive systems that do not rely on the presence of a expert engineer to keep them going.

The intermittent nature of the wind is clearly one problem that cannot be overcome by blade design or fancy controls. You either have to build windmills with a generating capacity that exceeds peak requirements so that surplus power can be stored, or you have to build equally over-large windmill capacity, but on such a large generating network that you can guarantee that there will always be enough wind blowing somewhere in the system to meet the demand. Whatever option seems most attractive, it will be almost impossible to devise a system that never experiences a shortage of wind power. A windmill that is not part of a large system can store power in electrical storage batteries, but these are expensive and relatively short-lived. A better system might be to split water into hydrogen and oxygen by electrolysis. The two gases could then be stored for subsequent use in a fuel cell, for example, which would consume the two fuels to generate electricity. Such a system would inevitably have its conversion losses at each stage, but the wind is a free fuel so the economics of the system depend upon the capital cost rather than the running cost. Yet another energy storage system for wind power could be the electric car. A windmill on top of the garage might be able to generate enough power to run a small electric vehicle.

Wind power would not be without its environmental impact. A fast windmill might be a noisy machine. And a community dotted with windmills might seem romantic in a Dutch painting, but when the windmill has been stripped of its glamour by commercialisation it will be recognised as just another eyesore. Windmills will also take up space. The US may have plenty of room for windmills, but in Europe land is more valuable. It might, however, be possible to build windmills on land and continue to farm it at the same time. This is something that will have to be looked into in a wind-power R&D programme. Where land is scarce, there is little doubt that windmills will be competing for space. The sea offers one way out. In addition to wave power being a wind-harnessing system, there is the possibility of building windmills at sea. According to the CEGB: 'if no other sources of electricity were acceptable or available, or if the price of electricity were to rise considerably relative to other prices, then a large proportion of Britain's electricity requirements could be supplied by windmills located offshore and coupled to a large energy storage system'.

Costs will ultimately decide the fate of wind power. 'I am convinced', said Professor Bergey, 'that large-scale wind power generating systems can be built for about $200 (equivalent 1974 dollars) per installed kilowatt.' This compares favourably with the cost of nuclear power (see chapter 6). The CEGB has different numbers. 'Costs of machines currently available in the UK range from £245 for a complete 200 watt installation to £3400 for a 1.125 kW set without batteries. These costs could possibly be reduced to the order of £500/kW by mass production but even so, with an installed capacity of 3–5 kW per household required to provide an average output of 1 kW (depending upon local wind speed), they would not compete with the grid supply or with a diesel powered generator at about £100/kW.' Conventional economic wisdom is unkind to something like wind power. Systems with consumable and expensive fuels may be cheaper to install, but how would they fare if their lifetime fuel requirements had to be paid for when the system were built? That is almost what you do when you buy a windmill or a solar power system.

Chapter 10

Using it wisely

There are many more ways of using energy than there are of generating it. Thus, while it may be difficult to assemble a coherent R & D programme on energy production, it is even harder to put together a comprehensive consumption R & D package. The whole area of energy utilisation R & D has been almost ignored until very recently. Thanks to the low price of energy – a price that actually fell in the 1950s and 1960s – we have not been very careful in our use of energy. Inefficiency in energy consumption abounds; it has been cheaper to waste energy than to spend the money needed to save it. Except in those countries where the climate is poor, houses have been built with insulation standards that are way below what can be achieved with a modest increase in capital investment. This is just one area where very little R & D is needed to achieve dramatic savings. New technology is not necessary, but people do have to be told the facts so that they can act upon them. Energy conservation depends upon the actions of every energy consumer, whereas energy production is in the hands of a far more limited clique of companies and governments and their agencies.

A breakdown of energy consumption by final users (see chapter 1) is an instant guide to where energy conservation can have the greatest impact. The domestic sector takes about a quarter the energy consumed in the UK, and about a fifth of the US's energy consumption. Much of this 'home energy' is for space heating. In the UK something like 70 per cent of domestic energy is used to heat houses and dwellings. Thus improved house insulation can make significant savings in this sector, which accounts for 13 per cent of Britain's primary energy consumption, that is, its consumption of the primary fuels, coal, oil and gas. As energy prices rise it makes more and more sense to invest capital to reduce running costs. This is true for all energy consumers.

Beyond this 'first order' energy conservation, which boils down to common sense and a realisation of the changed facts of life in the world's new energy situation, sits an almost unexplored area of R&D on energy use. This includes research on the efficiency of utilisation, which depends upon the development of new, less wasteful technologies to fulfil society's needs. It also encompasses research on the conversion of energy from its 'natural' forms into useful energy. The most obvious area here is the conversion of fuel such as coal into electricity. Because this is an instantly visible and sizeable use of energy, it has been looked into more thoroughly than other energy consumption technologies.

Future demand

The amount of energy we can save by being less wasteful depends on the enthusiasm with which we pursue conservation policies. The OECD's *Energy prospects to 1985* says that 'the estimated total potential for energy conservation (without recourse to emergency demand restraint) in OECD countries is 15 to 20 per cent of consumption levels previously forecast for 1985'. This implies a slower growth in energy consumption than in the recent past. The OECD's projections indicate that 'if there is no substantial change in the *real* prices of imported crude oil from their end-1974 level, then the overall energy consumption would grow at an annual rate of $3\frac{1}{2}$–4 per cent up to 1985 compared to the 5 per cent expected prior to the recent large increases in oil price'.

These savings are in line with most other projections of what a concerted and extended energy conservation effort could achieve. The savings will come about as a result of action in all consuming sectors. At the domestic level savings are most easily made by improving insulation standards; but it will take time to realise the full potential of such a programme. The place to start is in new buildings. (In many older buildings better insulation will not save energy, but it will make life more comfortable for the inhabitants who have probably been living in a house that is too cold anyway.) According to the OECD study: 'Assuming all new buildings are subject to tightened standards as from 1975 and making some allowances for increased insulation in existing buildings, the savings by 1985 could be 15–20 per cent of the

7   K P E

space-conditioning load, or 2.5–4 per cent of total primary energy consumption.' This may not seem a lot, but if the UK could achieve such savings today it would wipe something like £200 million off its oil import bill. Other savings can be made by lowering temperatures a degree of two – a temperature reduction of 1 °C saves about 6 per cent of the energy used for space heating. Lower light levels would also save energy, as would a switch to fluorescent lights which turn electricity into light more efficiently.

Fuel for mobility

Transport, which takes as much as a quarter of the energy consumed in some countries, is another easily recognised target for energy conservation. And the motivation to save energy is high here because this sector relies almost entirely on oil. The higher price of petrol at the pumps is a great incentive to drive more slowly, buy a smaller car, or even go by train. The US and Canada in particular could slash petrol consumption by building vehicles with engines of the sort of size that is more common in Europe. But a really sudden switch to 'compact' cars in North America could be very disruptive in the motor industry, and any energy savings would be reduced by the investment (both as money and energy) that would be required to set up new production lines. (These 'roll on' effects will apply to many energy conservation moves.) Beyond these changes, with their minimal content of new technology, new types of engine, for example, could save petrol. And we do not have to embark upon really 'way out' changes to save energy. The CPRS report *Energy conservation* (see chapter 3) pointed out that: 'for an equal power output, diesel engines are on average about 30 per cent more economical in terms of fuel consumption than petrol engines'. The novel engine that this report describes as 'the most attractive method of achieving a significant economy on the petrol engine' is the weak-mixture petrol engine, which burns a dilute mixture of petrol in air. The favourite version of this type of engine with Japanese car makers is the stratified charge engine, which might achieve fuel savings of 10 to 25 per cent but with performance and cost penalties. Steam turbines, Stirling engines, and other designs will require significant development before we can confidently predict how much fuel they can save, and what it might cost to achieve those savings. It will then take the motor industry a long time to gear itself up to producing new car models built around these propulsion units.

The fuel consumption of vehicles could be reduced without going as far as this. If car designers threw aside their pursuit of 'fashionable' vehicle shapes and decided to opt for aerodynamically efficient body shapes they could cut fuel consumption. Tests conducted by Britain's Motor Industry Research Association have shown that petrol consumption at 70 m.p.h. could be cut by 45 per cent if cars were properly designed. Car makers could go some way toward this target simply by rounding off the corners of vehicles, and by adding vanes and spoilers to existing bodies. The motor manufacturers have shown little interest in such ideas in the past: they have preferred to design cars for 'the public taste', rather than to try to educate the public to choose petrol saving shapes.

Aerodynamics are less important at slow speeds than the 'rolling resistance' of tyres on the road. 'Over 90 per cent of this resistance', says the CPRS report, 'is provided by the loaded tyres of the vehicle. It varies to some extent with the degree of inflation of the tyre and the materials of which it is made. Newer tyre designs, in particular steel-belted radial ply tyres, not only provide a substantial reduction in rolling resistance (of the order of 40 per cent) but also offer significantly longer tread life and improved handling of the car.' Britain has already taken to radial tyres, but motorists in the US have been slower to adopt them.

Along with other small changes, the modifications outlined here could help stretch the world's oil reserves. If (perhaps it should be when) oil does run out some other fuel will be needed to maintain our mobility. Artificial oil made from coal is one possibility (see chapter 4) which would require an R&D programme to assess the impact of new fuels on various engine designs and to ensure they are compatible. While artificial oil may be the easiest way of maintaining our ability to travel by air, there are other ways of powering ground vehicles.

Electric vehicles are not new, but their use as delivery vehicles and on confined sites where they can be recharged as necessary does not match the demands of the average car owner. Electric vehicles are too slow and they have a limited range because today's commercial lead–acid batteries are heavy and have a limited energy storage capacity. The sodium–sulphur battery is an attractive alternative to the lead–acid battery. In Britain this battery is being developed by Chloride Silent Power Ltd – a company created by the Electricity Council and the Chloride Group Ltd (which makes batteries). Chloride Silent Power was set up to take the sodium–sulphur cell developed in the R & D laboratories of the Electricity Council, and to develop it for full-scale production. The new company, which was set up in 1974, expected to spend more than £2 million on the project, which may not result in a marketable device before 1980.

An electric vehicle does not have to take its power from a re-chargeable battery, there are at least two other options. One alternative is a hybrid system with batteries and a small internal combustion engine. Such an arrangement could improve an electric car's acceleration and top speed, which are poor in existing electric vehicles, and give it the longer range that is essential if motorists are to accept electric cars. A second alternative to the rechargeable battery is the fuel cell, which generates electricity by direct con-version of chemical energy into electricity. A vehicle powered by fuel cells would carry a tank of fuel in the same way that today's cars carry petrol tanks. The fuel might be hydrogen, which would yield electricity in a process that is not dissimilar to electro-lysis in reverse. Despite several years of development, some of it as part of the American space programme, the fuel cell is not ready for true mass production, although it is suitable for some specialist applications, powering space craft for example. The hydrogen fuel cell is not without its drawbacks. Fuel cells are heavy and expensive, and hydrogen is a potentially explosive gas.

Even if fuel cells prove unsuitable for transport purposes, they could be significant in an energy system that has a role for a con-venient energy storage scheme that can produce 'instant' elec-tricity. A fuel cell could operate as a power station, which clearly places less importance on the weight of the system. If hydrogen does seep into our energy economy it will have to be made, and this would entail consuming some other form of energy. (Hydro-gen is not an alternative energy system – it is a way of storing and transporting energy.) Hydrogen might be made in a nuclear power station (see chapter 6) or by a wind or wave system (see chapter 9). Linked to a system of fuel cells, hydrogen could be turned into a convenient way of 'storing' electricity.

Returning to transport, yet another way of reducing the fuel consumed in transport is to encourage people to use public transport, which uses energy more efficiently than private vehicles. Such a move depends more upon social acceptance than on any R & D activity. The car owner now values his personal mobility so much that he will pay plenty for the privilege of running his own car, even when public transport is available. This brings home the sociopolitical nature of any campaign to reduce energy consumption.

Industrial consumption

The sizeable task of improving the efficiency of energy use in transport is relatively minor when set alongside the huge problem of increasing industry's energy efficiency. Industry consumes energy in countless ways, and the inhomogeneity of this sector makes it far harder to pinpoint key areas where energy can be saved. You cannot save energy if you do not know where it is being used in the first place. This inescapable fact of life hit Britain's policy makers when they tried to come up with energy saving policies for industry. The government found it easy enough to introduce legislation to raise the standard of insulation required in new homes, and it was little trouble to lower speed limits on the roads, but time and again the inadequacy of the information on industrial energy consumption made it impossible to act vigorously in this sector. And yet in Britain industry consumes around 40 per cent of the country's primary energy.

The problems of improving energy efficiency in industry are not solely technical. Industry could save significant amounts of energy by implementing techniques that are already well known. The general belief is that industry could save up to 20 per cent of the energy it uses by 'better housekeeping' – a phrase that cropped up again and again when a parliamentary committee was looking into the subject in 1975 – and by adopting the sort of heat insulation standards that are recommended for householders. Higher oil prices have reinforced the wisdom of saving energy to save money. Whereas energy is still a minor part of running costs for many companies, big organisations have such large fuel bills that it is only too obvious that they can improve their profitability by saving energy. The NEDO report *Energy conservation in the United Kingdom* spelt out the sort of problem that industry has to overcome before it even realises that it can save energy. 'The first requirement for energy conservation in industry is awareness by management that energy saving may be technically feasible and could give an economic return on the capital employed. In industries where energy intensiveness is low, there may be no detailed energy accounting and the assessment of potential savings might require a special study, but the energy difficulties of 1973/74 can have left few managements unaware of the problem.' Unfortunately those very energy difficulties brought with them such severe financial problems that many companies could not afford to save energy; they did not have the money to invest in equipment to cut energy consumption.

Some industries have a greater potential for energy saving than others. The iron and steel industry, for example, takes about 10 per cent of Britain's energy. Because energy is so important to the industry it has always pursued energy conservation more vigorously than less energy intensive industries. The efficiency with which energy is used in blast furnaces, for example, improved by 30 per cent between 1952 and 1962. The NEDO report predicted a further 20 per cent improvement in the next 10 to 20 years. Japan's Nippon Steel Corporation responded to higher fuel prices – which hit Japan harder than most countries because of its high dependence on imported fuel – by embarking on a campaign to cut by 10 per cent the energy that goes into making a tonne of steel. The company admits that 'the achievement of a further 10 per cent energy savings in what is already a highly efficient operation is a difficult goal that calls for substantial new investment and new technology'. The company identified blast furnaces as the easiest place to save energy but 'There are also possibilities in the development of ways to recover energy from red coke, waste heat from hot stoves, and top pressure of blast furnaces – areas that have been generally ignored to date because of the low return on investment.'

If the iron and steel industry wants to make a truly dramatic reduction in its consumption of fossil fuels it will have to abandon its traditional energy inputs and reducing agents. Nuclear reactors offer ways of meeting both needs. A high temperature gas-cooled reactor (see chapter 6) could provide both heat and electricity; it could also generate hydrogen which could replace carbon as a re-ducing agent for turning iron ore into iron. There are several ways of running an iron and steel industry on nuclear power. In 1973 the European Nuclear Steelmaking Club was formed to study this subject. As yet the club – which brings together steel-makers in Belgium, France, Germany, the UK, Italy, Luxembourg, and the Netherlands – has restricted itself to paper studies of the most promising avenues of R&D. The club's first significant conclu-sion was that the first nuclear steelmaking complex built in Europe should have the reactor and the steel plant located separately. The reactor would not supply heat and hydrogen directly to the steel plant because such a close relationship could cause problems when the reactor was inoperative, which is bound to be a more common event than a shutdown at the steel works.

Nuclear steel plants probably will not be with us before the end of the century. More immediate improvements in the energy efficiency of the steel industry will rely on less exotic develop-ments; but even here there will be a delay before more efficient technologies can be taken up. A report from the Dutch Future Shape of Technology Foundation, *Energy conservation: ways and means*, concluded that 'little or no energy conservation can be ex-pected from new processes between now and 1985. Quite apart from the fact that there is little incentive to develop new steelmaking processes in view of the current high efficiency of the blast furnace/ oxygen steel combination, there are virtually no energy savings to be gained from the processes being developed at the moment.'

The Dutch report, which is one of the most detailed studies of what could be achieved with energy conservation technology, has tabulated an analysis of the 'potential for savings in industrial energy consumption' for some of the most important industries in the Netherlands. The OECD report *Energy prospects to 1985* predicts that, whereas 'industrial energy conservation is limited in the short term by the present inventory of equipment and processing systems...more investment in conservation coupled with technical innovation could increase the sectoral savings to

15 per cent of previously expected consumption levels by 1985'. In the short term, industry could cut its fuel consumption by 10 per cent and its electricity consumption by 5 per cent 'without seriously affecting output'. Industry could also save energy by recycling materials, which often can be processed with less energy than fresh raw materials. As with radical changes to cut energy consumption in the transport sector, far-reaching alterations in the 'lifestyle' of industry will not be without drawbacks. The greater the obstacles, the longer any change will take. Organisational and political changes, as opposed to progress with R & D, could speed up conservation. However, responsible energy use is not encouraged by the growing business practice of passing rises in energy costs (or any other material input) straight on to the customer with almost no effort to improve efficiency. Governments can help by, for example, putting industry on the spot by asking companies to put some sort of energy 'account' into their annual reports. This was one of the moves adopted by the British government when it produced its first energy conservation package at the end of 1974.

Heat and power

Energy conservation is not simply a case of using what energy is available with greater efficiency. Just as the energy input needed to produce a particular level of industrial output can be cut, so can the crude energy input that goes into producing a useful energy output. The energy 'wasted' in the generation of electricity is a natural consequence of our need for high grade energy. This waste arouses violent emotions among some energy watchers. More extreme conservationists look at the energy content of the fuel that goes into a power station; they see that as much as three-quarters of this input energy ends up as hot water and warm air that is no good to anyone. Just a quarter of the energy ends up as useful electricity. There are two ways of reducing this energy waste: we can either improve the efficiency of conversion from fuel to electricity, or we can put the waste heat to good use. The first course depends upon technological progress and the development of new conversion technology. Waste heat utilisation already is practised to a limited extent.

District heating is one way to put the hot water that comes from a power station's turbines to good use. A district heating system takes this thermal energy (hot water) and puts it into a network that carries it to industrial and domestic premises for space heating. Space heating is a major drain on energy consumption in the developed countries, so we can achieve significant savings if some of that demand can be met by energy that would otherwise go to waste. Power stations have to operate their turbines differently to obtain water that is hot enough to be useful in a district heating network; and they lose some electrical output in the process. The steam that drives the generating turbines in a combined heat/power system is not exhausted from the turbines at atmospheric pressure as is the usual practice, but at a higher pressure so that it can be condensed at a temperature above 100 °C. This 'back pressure' operation lowers the efficiency of electricity generation; a back-pressure turbine produces 15 to 20 per cent less electricity than a conventional generator.

While an electricity utility might not welcome a drop in its generating efficiency, if overall efficiency in energy use is to be boosted then district heating has its attractions. In principle, district heating combined with power generation can put to use 85 per cent of the energy in the power station's fuel, but only if the conditions are right. The demand for heat and electricity have to be at the right level, and they must not fluctuate too much. Life is never that simple. According to the NEDO report *Energy conservation in the United Kingdom*: 'The wide fluctuations in the demand for heat in housing are likely to lead to imbalance between electricity generation and the demand for steam, so that an average thermal efficiency of 45 per cent is more likely over a full year's operation than the theoretical 75 to 85 per cent that might be achieved under ideal conditions.' At this level of thermal efficiency it might be better to build two completely separate systems: a power generating network, and a separate heat generating system linked to a district heating network.

The massive task of plumbing Britain's towns into a district heating network would be unbelievably costly. It might be feasible to put district heating into a new town built on a 'green field' site, for example. And industry can sometimes make good use of a 'total energy scheme' which generates both electricity and steam for process heat. Factories often generate their own steam while they import electricity from the local utility. A total energy system on the site could meet both needs, but there can be institutional problems in the way of such schemes. For example, a factory with a total energy system might want to buy or sell electricity. In Britain the utilities have a reputation for discouraging such transactions by 'rigging' the tariffs against those who do not want a regular and predictable supply of electricity.

Sweden started installing district heating after the last war. In the city of Vasteras, for example, 98 per cent of the town's 47 000 dwelling units (flats and houses) are on the district heating network. In Sweden the various local district heating schemes are fed both by heat-only boilers, and by combined heat and power units. In this way it has been possible to take district heating into small areas. As the area covered by these smaller boilers grows, it becomes possible to connect them to a central power/heat station. The Swedes hope to be able to develop nuclear power stations as sources of both electricity and heat. Before that can happen, the country's engineers will have to develop systems that make it possible to transmit hot water over long distances without excessive loss of heat, and that do not cost too much to install. The Swedes are studying plans to establish three nuclear heat/electricity schemes around Stockholm, Gothenburg, and Lund. Reactors would probably supply base-load heat, with local fossil-fuelled units meeting peak loads.

District heating arouses passions of surprising intensity. The sight of huge cooling towers alongside a power station makes the advocates of district heating see red. They look upon the waste of heat on such a scale as a crime; they do not count the cost of putting that heat to good use. On the other hand electricity utilities see their job as that of supplying electricity, not heat. By supplying heat they might do themselves out of some of their market for electricity. This narrow-minded 'electricity-only' thinking suddenly looked sillier than usual after the energy upheavals of 1973/74. In Britain, for example, there was considerable pressure on the Central Electricity Generating Board to review its policy on combined heat and power generation. The CEGB, much abused for its attitude to district heating, reviewed the use of reject heat from power stations in its reassessment of unconventional energy systems. It concluded that 'with the rapid increase in fossil fuel prices, simple district heating schemes using large central boilers are more attractive than hitherto, particularly for compact new developments, such as redeveloped city centres. However, in assessing the overall benefits, the time to build up the load, and the inconvenience and costs of digging up roads and scrapping domestic installations in existing communities need to be taken into account.' The CEGB came to the conclusion that 'because of problems of matching the heat load and supply, it is likely to be more economic to first install simple district heating schemes, and only consider connecting them to power stations when a sufficient load has been built up'.

Waste heat could be put to use now: 'There would be scope for reducing the cost of heating glass houses by siting them at power stations to make use of reject heat', says the CEGB. At the time of the rapid rise in oil prices, the impact on horticulture caused especial concern because 90 per cent of the UK's glass houses are heated by oil. The CEGB estimated that 'the heat rejected from a single 2000 MW power station is enough to heat all the 3000 acres of heated glass houses currently used in England and Wales for tomato growing'. Waste cooling water from power stations is not normally hot enough to warm a glass house, but it might be possible to design a new type of glass house with more efficient heat transfer. Another possibility is direct heating of agricultural ground with power station waste heat – either by sending hot water through underground pipes or by spraying it on to the soil. Fish farms could improve their efficiency by heating fish ponds with power station cooling water.

The CEGB also came to the conclusion that nuclear power stations are potentially valuable suppliers of heat as well as electricity; 'Use of reject heat from steam turbines will become more attractive when there is much more nuclear plant installed. Use of nuclear stations to provide cheap process steam for industry could then be particularly attractive.'

Few of these energy saving ideas depend upon a significant R & D programme. More often than not all we need to do is review the existing techniques in the light of the new circumstances. The CEGB study came to the conclusion that 'there appears to be no need for further research on district heating since techniques are

▶ More efficient ways of generating electricity would extend the life of the world's fuels. One way of achieving this goal is to employ magnetohydrodynamic generators. The Soviet Union has the most advanced MHD systems and is cooperating on MHD research with other countries, including the US. The photograph shows the Y-25 MHD power station in Moscow.

well established. There could, however, be a need for research on use of reject heat for horticulture and fish farming.'

Magnetohydrodynamics

An improvement in the efficiency with which power stations turn fuel energy into electrical energy depends on the development of new conversion technologies, and this depends on a significant R & D effort. Over the years there has been a steady increase in the efficiency with which fossil fuel is converted into electrical energy – from about 20 per cent in the 1940s to 30 per cent or so in the mid-1970s – but a truly significant leap in efficiency can only come about with new techniques and technology. Possibly the best prospects are magnetohydrodynamic generation (MHD), and power generation using superconducting technology. (Superconductors could also improve the efficiency with which electricity is transmitted from the power station to the consumer.) Both MHD and superconductors have been supported by significant R & D programmes – a support that has waxed and waned as fashion changed. The Soviet Union has made the greatest progress with MHD: a pilot plant started supplying electricity near Moscow in 1971.

MHD aims to rid electricity generation of the massive rotating machinery that is part of today's turbogenerator equipment, and to boost the efficiency of conversion to as high as 50–60 per cent. In an MHD generator a flow of hot ionised gas or a stream of liquid metal replaces the turbogenerator's rotating conductor. The conducting fluid flows through a magnetic field which induces a current flow in the plasma. Electrodes immersed in the conducting fluid can tap this current which is subsequently fed to the supply grid. The hot fluid can come from a variety of energy sources. There are three main types of MHD system – open-cycle plasma, closed-cycle plasma, and closed-cycle liquid metal.

In an open-cycle MHD generator fuel is burnt to yield hot combustion gases. A 'seed' is added to increase the electrical conductivity of the gas which then goes through a nozzle to speed it up before it flows along the MHD generating channel with its magnetic field and current collecting electrodes. The fuel combustion system can be straightforward if the fuel is a gas or an equally clean-burning fuel. But coal will be the mainstay of tomorrow's fossil-fuelled power stations, and coal combustion is a messy business generating a variety of abrasive and corrosive combustion products. There are several options for a coal-burning MHD system. A study by the US Bureau of Mines lists four alternatives: single-stage combustion, suspension gasification with air, cyclone gasification–combustion, and suspension gasification with hot flue gases. Single-stage combustion systems would be the easiest to build, but the abrasive particles produced when coal burns would be easily carried over into the MHD system and would exacerbate the already serious problem of MHD channel corrosion. The corrosiveness of coal's combustion products can be reduced if the coal is gasified to produce low-Btu, hydrogen-free gas that is cleaned before combustion in an MHD device. Such a system would be more complex and expensive than a simple single-stage combustion system.

A group set up in the US by the National Petroleum Council (NPC) reported that far from it being a simple system (which MHD is in theory) the need 'to add and recover seed, preheat air to high temperatures and to pass potentially damaging seed material through steam plants as well as the possible need to recover sulphuric acid and nitric acid...moves MHD from a relatively simple concept to a complex one'. Thus the prospects are not good for an early introduction of this technology. The OECD's *Energy prospects to 1985* concludes that new generating systems, such as MHD, 'show little if any promise for practical application over the next two decades'. The OECD's energy R & D report that was published soon after this study said: 'One of the basic problems still to be resolved is that of the useful life of the materials from which the various components of MHD generators are made; it is therefore likely that in the first place MHD will be used for peak power generation.' The CPRS *Energy conservation* report was so pessimistic about the prospects for MHD that it ruled out any British programme on the subject; 'Despite the continuing interest in MHD in the United States and the Soviet Union, and the sizeable effort put into it by the CEGB in the early 1960s, the case for doing further research is weak.' The NPC report pointed out that few of the people who would have to buy MHD systems to make the technology a success show much interest in the idea. 'No recognisable widespread interest in the MHD concept for electricity power production exists in the electrical manufacturing industry; the coal, oil or gas industry; the electric utility industry or government.' The NPC is not too sure that technical success is within easy reach: 'There is some probability that attainable engineering solutions [to technical and engineering problems] will not be economically or practically acceptable.'

This assessment applies principally to open-cycle MHD plants fired by fossil fuels. Even more exotic concepts, closed-cycle plasma generators and liquid-metal generators, are further in the future. These will be matched to nuclear reactors if they are developed at all. A high temperature reactor with an outlet temperature of 2000 °K could achieve a generation efficiency of between 50 and 55 per cent. A liquid-metal generator could operate at lower temperatures (down to about 900 °K) and might, in tandem with a steam turbine, achieve an efficiency of 45 to 50 per cent. As yet these are strictly theoretical targets.

The electricity supply industry does not need technologies as adventurous as MHD to boost its efficiency. Gas turbines have been improved and operated at higher temperatures as new materials have been developed for turbine blades. Gas turbines are now employed as electricity generating units, but they burn expensive high quality fuels so their use is limited to peak power generation. They are ideal for this purpose because they are inexpensive to build and can be brought into operation very quickly. Gas turbines might become attractive for base load generation if nuclear reactors could be made to produce hot gas that could be exhausted through a gas turbine. If the hot gas is subsequently used in a steam turbine cycle to recover more of its energy, the conversion efficiency of generation can be 50 per cent or higher.

Superconductors

Somewhere between the long-term prospect of MHD and the relatively modest step forward offered by gas turbines there are the developments that require some R & D to match new technology with established techniques. Superconducting generators, for example, could increase conversion efficiency by reducing the size and weight of generating machines. Power transmission could be improved by superconducting technology; and superconductors might make it possible to store electricity.

Superconducting materials lose their electrical resistance at very low temperatures (below 20 °K, although this is rising as new superconductors are found). Superconductors are better suited to direct current operation, so generator design poses a few problems. Small alternating current and direct current machines have proved that superconducting generators can be built without too much difficulty. In the US Westinghouse has operated a 5 MW AC machine; and in the UK a 2.4 MW DC machine has operated at the Fawley power station. This limited experience and various theoretical studies have shown that a superconducting generator would be about the tenth of the size of a conventional machine of the same generating capacity.

The potential market for superconductors makes it worthwhile to develop this technology and to overcome the problems that go with it, such as the difficulty of cooling a superconductor below 20 °K and keeping it cold. MHD work was a spur to superconducting magnet development in the UK before the CEGB shut down its MHD work. Superconducting magnets have, so far, found their largest market in the accelerators built by high energy physicists. And there will be a new interest in superconductors as the fusion programme moves into the next phase of development (see chapter 7).

Superconducting power transmission looks like the simplest application of this new technology. And the motivation to develop this use of superconductors is clear. Overhead transmission of electrical power loses between 6 and 10 per cent of the power carried; superconducting transmission lines would lose about 1 per cent of the power carried. As an added bonus, underground power transmission would be easier with superconductors, making it possible to bury those unsightly transmission lines. Clearly superconducting transmission would add a new factor to the cost of electricity transmission, the cost of cooling the lines. According to the CEGB: 'Over half the cost is associated with cryogenics, including about 10 per cent for helium.' (Helium is the most suitable cooling fluid for superconducting materials.) The CEGB's assessment of the development of superconducting cables is that 'with an energetic programme construction of a prototype cable could start in about 1980'.

The CEGB is less optimistic about the prospects for superconducting energy storage: 'the scale and economics do not invite vigorous development'. More visionary souls are spurred on by the prospects of an energy storage system of such power that 'a superconducting system the size of a football field could act as a peak-shaving unit for a large metropolitan area, such as Cleveland'. William Keller, of the Los Alamos Scientific Laboratory, described superconducting storage in this way; he added that energy storage could be important: 'largely because of daily, weekly and seasonal variations in the demand by consumers for electric power, we are using only about 60 per cent of our power generating capability. This could be remedied if we could store energy during slack demand periods to be delivered during peak demand periods.' The concept of superconducting electricity storage is simplicity itself. AC power from a generator is converted to DC power which flows to a superconducting magnet which, because it is superconducting, consumes no current; to store more energy in the system, the current is simply increased. The electricity is recovered by the reverse process. One drawback would be the physical effort needed to hold the magnets together; in the same way, the large magnet in a fusion reactor would exert a strong repulsive force that would have to be restrained if the magnets are not going to fly apart. This problem could be even more severe with a superconducting electricity store, which exists solely to hoard energy and would have to be designed to achieve the strongest field possible. One way of overcoming this difficulty would be to embed magnets in rock that could hold the whole thing together. A research project on this idea, at the University of Wisconsin, has come up with an estimate showing that 'such devices in large scale, possibly 50 metres radius or 100 metres radius would be sufficiently inexpensive to be of interest to a power system'.

Electromagnetic energy storage might be about 95 per cent efficient, but the cost would be high. The stage of development of today's superconducting technology makes it difficult to be precise when estimating the likely costs of electromagnetic energy storage. The CEGB has extrapolated the sort of cost figures that are being talked about among fusion engineers to arrive at a tentative estimate of the cost of superconducting storage. It arrived at a figure of £1000 per kWh of stored energy. As well as cost, there are some significant technical problems standing in the way of superconducting electromagnetic energy storage: high-voltage insulation for cryogenic systems; high-current leads into the very low temperature of the storage area; and the control of the electrical flow to and from the store. The CEGB says, 'even the most optimistic assessments do not envisage the system breaking even with other storage possibilities below about 10 GW sizes and, for this reason, the system cannot be given high priority'.

A system that can retain energy generated when demand is low, for subsequent consumption when it is higher, can reduce the generating capacity that a utility has to install; or it can keep the most economical power stations in operation more of the time, reducing the need to run the less efficient and more costly plants. Today's pumped storage hydroelectric schemes fulfil this role to a limited extent. In future this capability will become more valuable as nuclear reactors make sizeable inroads into the electricity supply system. Nuclear power stations are costly to build, but inexpensive to run. Hence it is easier to run reactors at a steady power supply level than to raise and lower their output to follow the fluctuating demand. No country has yet reached a state of affairs where it has to reduce output from nuclear power stations during periods of low electricity demand but it could begin to happen in the 1980s if today's plans to expand nuclear capacity are realised.

The economic viability of energy storage depends on the efficiency with which energy can be stored and recovered, the fraction of the system that is nuclear capacity, the price of fuels, and the shape of the electricity demand curve. If you cannot smooth out the supply of power by storing it, then you have a limited ability to switch the demand for electricity away from the natural peaks. Britain has done much to smooth the troughs and peaks of demand. Throughout the 1960s the UK electricity supply industry encouraged 'off-peak' heating, by introducing a system of electricity tariffs that made it economical to install space heating systems that took in night-time electricity and stored it as heat for subsequent use.

Electricity is not as easily stored as low grade heat. We have already looked at one advanced system in the context of super-conducting systems. Most of the more conventional energy storage schemes are variants of pumped storage. Sites suitable for hydro-electric pumped storage schemes are few and far between, and in any case there is always violent opposition from environmentalists whenever a utility puts forward a new pumped storage scheme. Hence there is growing interest in underground pumped storage schemes. These can be simple pumped hydro schemes with one water reservoir underground, and the sea or a river acting as the second reservoir. There is not much support for this idea; compressed air pumped storage looks more promising. A compressed air storage scheme takes excess power during periods of low demand and uses it to pump air into an underground reservoir. When the energy is to be retrieved it is used for combustion of high grade fuel in a gas turbine. Pumped air storage would be more efficient than a simple gas turbine in turning fuel energy into electricity. In conventional operation a gas turbine has to use up some of its energy to drive an air compressor. If the air is compressed using energy from another source, a nuclear power station for example, a gas turbine can be run without a compressor stage. In this way the turbine fuel can yield something like three times as much energy when burned in a compressed air storage system. The West German electricity utility Nordwestdeutsche Kraftwerke is building a 290 MW installation which should start operation in 1977. This scheme will generate 580 MWh of electricity during peak periods; it will consume 930 MWh of gas turbine fuel and 480 MWh of off-peak electricity. An ordinary gas turbine would use 2 to $2\frac{1}{2}$ times as much gas turbine fuel for the same output.

Other energy storage options include batteries, fuel cells, and even giant flywheels. These last might be made of a composite material of some sort, which, because of its strength, could be used for a flywheel which could store more energy than a flywheel made of a heavier material such as steel. Price is the obstacle to flywheel development; none of the materials that might fill the bill are cheap enough to bring flywheel energy storage anywhere near commercial reality. Carbon fibres, for example, would, at today's prices, result in an energy store that is an order of magnitude dearer

▶ Energy is not easily stored, and large amounts of electricity are impossible to store with present-day technology. Pumped hydroelectric storage is one of the few ways of storing electricity generated at one time for use later on. And any pumped storage scheme worth doing these days is a very large undertaking.

At Dinorwic in Wales the Central Electricity Generating Board is building one of the world's largest pumped storage schemes. This will take electricity generated in off-peak demand periods and use it to pump water into a high-level reservoir from a low-level lake. The stored energy can be recovered in the same way that a hydroelectric scheme works – water flows from the upper to the lower reservoir, driving large turbines.

The Dinorwic project will be able to generate up to 1500 MW of electricity for around $5\frac{1}{4}$ hours. A major function of the scheme will be as 'spinning reserve' – that is generating capacity that can be turned on in seconds to take over if a huge generator breaks down somewhere else in the system.

The photograph shows the size of the civil engineering works at Dinorwic, where a huge tunnel is being bored between the two reservoirs. Before the £100 million project got going, there was a major battle between the environmentalists and the CEGB.

than anything an electricity utility could afford to pay. There is no point in storing electrical energy if the stored energy costs a lot more than electrical energy generated by burning fuel.

R & D on new batteries has always offered significant rewards to the company that could steal a march on its rivals and come up with a cheaper and more efficient device. However, the motivation for battery development has not been the desire to produce energy storage systems, but the need for portable power units. The OECD report *Energy prospects to 1985* summarises the status of battery research:

'The progress made during the last few years justifies a sustained R & D effort and suggests that there will be a steady development during the 1980s. Batteries which seem of particular interest are zinc–air cells for low-temperature use and lithium–sulphur, sodium–sulphur and lithium–chlorine cells for high temperature applications (operating between 300 °C and 700 °C). As in the case of fuel cells, electrochemical and materials research is necessary, in particular with a view to increasing the useful life and the number of charge–discharge cycles of these batteries.'

Fuel cells are not too far removed from batteries in that both would benefit from R & D on electrochemicals and materials. Fuel cells have cropped up at several points in this journey through the energy R & D system: now is the time to take a closer look at them. A fuel cell turns the chemical energy of a fuel into electrical energy. The easiest cell to understand is the hydrogen fuel cell. The input to this cell is hydrogen and oxygen (from air); and the electricity production process is akin to electrolysis in reverse. Hydrogen and oxygen are continuously fed to the cell's anode and cathode respectively. Electrochemical reactions in the cell produce electricity and water. Fuel cells can be made to run on fuels other than hydrogen – alcohol, carbon monoxide, and hydrocarbons are all possible alternatives. A fuel cell can turn a fuel's chemical energy into electrical energy with an efficiency of something like 50 to 70 per cent. A fuel cell would be both clean (a hydrogen fuel cell would produce water as its major waste) and quiet.

The OECD's *Energy R & D* says: 'A substantial R & D effort would still seem to be necessary, particularly in electrochemistry and materials science, before fuel cells with sufficiently low capital costs, adequate working life and using relatively cheap fuels can be used for generating electricity other than in very special circumstances (for example in space applications).' R & D on fuel cells has been in and out of favour over the past 20 years. Early on there was a mad rush to enter this area, but the scramble died down when it became clear that fuel cells would not be an overnight sensation, and that they would be produced commercially only as a result of a sustained and expensive R & D effort. NASA paid for some of the early R & D: fuel cells were an invaluable tool in the space programme. But ultra-sophisticated aerospace technology is not the sort of thing you want to find in a power station or under the bonnet of a motor car. For example, fuel cells will not penetrate the energy market if they continue to rely on expensive catalysts.

Energy R & D says of the fuel cell's prospects: 'In view of the extent of the technical and economic problems yet to be resolved, it is unlikely that fuel cells will be applied on a large scale before the middle of the next decade.' An earlier study, carried out by the US National Petroleum Council's new energy forms task group, was even more gloomy: 'unless major new developments occur in fuel-cell technology (unlikely, considering the vast amount of work completed), fuel cells will have little impact on fuel utilization by the end of the century'. The NPC report is not really at all enthusiastic about fuel cells: 'In any case, the gain in fuel efficiency over conventional converters would be small (10–15 per cent)'; and this would not exactly wipe the floor with improvements we can expect with less tentative ideas.

WASH-1281 left nothing out of the repertoire when it told the president of the US to spend money on R & D; it called for a spending of $80 million on fuel cells over the 1975–80 period. It certainly makes sense to spend some money on fuel cells if you believe that the hydrogen economy could become a reality. The CPRS pointed out in *Energy conservation* that the hydrogen fuel cell 'presents some formidable technical problems'. The CPRS was looking at the potential impact of fuel cells on transport, which was unlikely to be great 'in the foreseeable future'. There may be a whole catalogue of problems facing the hydrogen economy but there is something of a bandwagon developing in support of the idea. (The bandwagon's momentum was hardly multiplied by ERDA's request for just \$$\frac{1}{2}$ million for fuel cell R & D in FY 1976, to match its similar expenditure the year before.)

Hydrogen

The hydrogen economy is another of those topics that crops up throughout any analysis of energy R & D. Hydrogen is not a form of energy: it could be a convenient way of storing and transmitting energy from place to place. In theory, hydrogen could fill most roles in the energy repertoire. Nuclear reactors and many other energy sources could produce hydrogen, even if they had to resort to the inefficient process of using electricity to produce hydrogen by electrolysis. The resultant fuel could then power our transport system. (And combustion of hydrogen in an internal combustion engine would produce none of the oxides of carbon or unburnt hydrocarbons that come out of a petrol engine; nitrogen oxides could, however, remain a problem.) Hydrogen could replace methane in the natural gas transmission and distribution network. Hydrogen could even remove the need for long distance electricity transmission if fuel cells were used to generate electricity at a local level from hydrogen piped over a transmission system. This versatility has seduced many researchers into commending hydrogen as a suitable topic for large R & D spending.

The first task in any hydrogen R & D programme would be the development of efficient and inexpensive ways of making hydrogen. Quite clearly nuclear reactors must be a favoured energy source for the hydrogen economy, but a reactor cannot reach the temperature required for direct dissociation of water into hydrogen and oxygen (more than 2500 °C). And electrolysis – the hydrogen production 'reference' against which all other techniques are judged – is very inefficient, thanks partly to the need to go through the intermediate electricity generation stage. Various research teams are looking into multi-stage chemical methods of breaking water to yield hydrogen. (The best known project is at Euratom's Ispra research centre in Italy.) These tentative programmes are far from a stage at which they could provide the basis for a fully fledged hydrogen industry. In the middle of 1974, Dr Keith Dawson, head of Britain's then newly established Energy Technology Support Unit, said that there were ten different research establishments looking into thermo-chemical methods of making hydrogen. So if hydrogen can be made in an acceptable industrial process, a reasonable path should not be too elusive. Dawson warned, however, that none of the cycles looked at so far (involving various series of chemical reactions) had shown much sign of being an outright winner.

If hydrogen can be made at a reasonable price it should be easy enough to send it from place to place in pipelines. *Energy R & D* says: 'The cost of transmission for hydrogen is estimated to be some two to three times higher than for natural gas but much lower than for electricity, especially over long distances. Storage is also two to three times more costly than for natural gas, but liquid hydrogen can be stored at a fraction of the cost of storing electrical energy in pumped-storage hydro schemes.' Storage of hydrogen is one area that is ripe for R & D, but only if it looks as if hydrogen stands a good chance of becoming an important fuel. Hydrogen is not the easiest thing to store, so it might prove worthwhile to develop systems such as metal hydride storage.

Unlike fusion, which few people dismiss as an irrelevant area of energy R & D even if they do feel that its impact will be very long term, the hydrogen economy has its opponents. The electricity industry believes, of course, that people will always need electricity, so why not use this as *the* energy carrier? Why bother, the 'electricians' say, to create a parallel energy transmission and distribution system just because the natural gas pipes are there? They see the existing natural gas network as something that would cost a great deal to save if hydrogen is the saviour. The gas industry says that the gas transmission and distribution system is such an established part of the energy economy that we will always want gas; and hydrogen is the most obvious long-term replacement for the natural gas that will soon start to run out. The coal industry has its line on this argument. A tank of hydrogen contains about half as much energy as an equal volume of methane (natural gas); so the coal men want to turn hydrogen into methane by gasifying coal. A nuclear reactor could provide the energy for the system if it is important to use the coal as a feedstock rather than as a source of energy at the gas production stage; so the nuclear industry does not have to miss out even if we decide not to 'hydrogenate' our gas system.

The debate on the hydrogen economy will tick over for quite a while as the hydrogen R&D programme progresses. It is important for this effort to continue because at the current state of knowledge we do not have enough information to say with any conviction whether or not the hydrogen economy is a starter or a waste of money. But the hydrogen R&D programme should not be allowed to go on to the final conclusion just because money has been invested in the idea of hydrogen production and distribution. We may come to regret it if we do not look closely at the concept of a hydrogen economy, but we could regret it even more if the ultimate benefits of this development are won at a crippling cost.

Most areas of energy R&D deserve equally close scrutiny as projects change from research possibilities to the development phase. Many of the world's R&D laboratories are now switching their efforts towards energy activities. Only a fraction of the projects now under way or about to be started can hope to win through to industrial implementation. It will require at least as much good judgement to stop a project as was needed to set it up. We must, therefore, constantly reassess the energy R&D programmes now under way or about to be launched to see if they fit in with social requirements, which will themselves change throughout the next 50 years. (This may seem like a long time to be worrying about, but much of the R&D that is now contemplated may not begin to make any impact before this period is over.) Any R&D project has a tendency to take on a life of its own. It will take strong management to ensure that we control our energy future, and to prevent the R&D projects from growing into monsters that no amount of opposition can stop.

Chapter 11

The energy equations

Science, technology, energy reserves, and the pattern of energy consumption are not the sole deciding factors in the formulation of an energy R&D programme. I have already mentioned money, water, and the 'social' constraints, such as environmental standards, that are also part of the energy R&D scene: there is a further factor that many people might consider less important but which looms over everything the research community tries to do. This final factor is the political input inserted into every aspect of life by those we elect (or who elect themselves) to control our destinies. It would not do to ignore this final input into the energy equation. In this chapter I will look briefly at this input into energy R&D before going on to look at the way in which the world's energy system could develop.

Energy policy receives political inputs that sometimes have nothing to do with energy technology. There may be questions about the safety of nuclear power, fo example; but a minister with significant power over the direction of a country's energy system may take decisions with very little understanding of the technical issues, and in some cases little or no knowledge of the relevant literature. Of course, there are advisors providing technical input to the decision makers, but some have not much more than a superficial knowledge of the subject (they can be about as knowledgeable as a competent technical journalist).

Let me give an example: the US is establishing a Solar Energy Research Institute (SERI). And as with all new institutes, one of the first jobs was to find a home for SERI. When President Ford hinted that certain states were on the short list of potential homes, politicians and technologists in unlisted states were very upset. They were upset because the criteria for a state appearing on the short list were political rather than technical. (This episode was just before an election, when energy 'consumers' suddenly became 'voters' and every decision's political content was amplified.) A similar conflict of interests between officialdom and technologists occurred in Europe when the EEC was seeking a home for its new fusion R&D project, JET (see p. 126). The EEC's officials favoured a site (Ispra in Italy) occupied by one of their own R&D laboratories. Most of the fusion scientists in Europe on the other hand wanted nothing to do with the site, which had an appalling history of industrial unrest and was universally thought of as scientifically second class.

Hidden influences

Political influences such as those described above are fairly easy to spot, especially when they are so obviously political that opposing politicians can make mileage out of the situation. There are, however, 'political' influences that are less easily detected. The Energy Policy Project of the Ford Foundation (see p. 40) is an example. When the EPP produced its final report, *A time to choose*, it was attacked for its errors in economic assumptions, among other things. An oil company went so far as to place advertisements in the *New York Times*; and in an 'anti-report', *No time to confuse*, published by the Institute for Contemporary Studies, Armen Alchian, Professor of Economics at the University of California, said:

'A famous economic principle of demand is ignored in the report: the principle that the amount of petroleum demanded depends on the *price* of petroleum. The lower the price the more we "need", require, or demand. That the report of a $4 million project would ignore such a well-established powerful fact of life would be incredible were it not that so many politicians, bureaucrats and even oil industry people also ignore or deny it.'

I have not gone into great detail about the economics of the energy business; but I have tried not to ignore this fact of life which Professor Alchian finds so obvious but which is too often ignored by energy researchers. This and other economic factors clearly influence energy R&D and will have an impact on the introduction of new energy technologies. The western world saw a dramatic reduction in energy consumption after the energy upheavals of 1973/4. Some of this reduction was undoubtedly due to the economic slump that was already under way when the Arabs

▶ The nuclear nightmare?

imposed their oil embargo and implemented their massive price rises, but the higher prices were themselves a factor in reducing consumption. Higher energy prices were a major driving force in industry's energy conservation initiatives – industry was not interested greatly in saving energy so that mankind could stretch out its valuable resources, but the incentive of higher prices and reduced profits was very strong.

As energy consumption fell the energy industry decided that some of its expansion plans could be shelved. In Britain, for example, the electricity supply industry lost something like a couple of years' growth. This allowed the government to embark on a much smaller nuclear power programme than might have been demanded had economic activity and electricity consumption been rising at historic rates. The slump also made it easier for the government to select a relatively unproved reactor design for the next stage of the country's nuclear programme. Had there been more urgency, the pressure to choose a proved design that could have been ordered in quantity and built immediately would have been very strong. The same economic slump forced down oil consumption and made the US energy industry think twice about plans to create new plants to extract oil from shale, for example. Lower energy demand was not the only factor causing these delays and second thoughts – shortages of cash and government dithering also had an impact, as did inflation which pushed up the cost of alternatives to natural gas and crude oil.

Higher oil prices will clearly depress the rise in energy consumption; and the rise that sets in as economic growth continues will be slower than in the past. Just what the relationship between gross national product (GNP) and energy consumption will be in future is anybody's guess. The prevailing price of oil will clearly be an important, if not determining factor. There is a school of thought which maintains that the oil exporting countries will slash the price of crude oil, either to push up consumption to increase their incomes, or to eliminate the alternatives to cheap Middle East crude – such as expensive offshore oil, or oil from shale or tar sands, or synthetic crude from coal. As economic activity rises so will energy consumption, and the oil exporting countries will soon find that they can sell as much oil as they need to maintain their development programmes. The oil producers' cartel may be fragile, but it has endured the most severe disruptive pressures – the low oil consumption between 1974 and 1976, for example. And there seems little sense in the oil exporters' selling oil at a low price merely to put Britain's North Sea oil operations out of business. (North Sea oil may mean a lot to the UK and Norway, but it is a small fraction of the world's energy supplies.) Another factor is worth remembering – some of the more perceptive oil producers have talked about the 'synthetic crude' business as a yardstick by which they can set their prices. The theory is that the price of 'natural' crude should be geared to the price of the competition.

Only the US, with its large home production of crude oil, could temporarily avoid the higher oil prices of the future. This might happen if the US's 'regulation' policy continues and the country continues to support an oil and gas market with prices of domestic supplies substantially below those of the world market. While energy consumers in the US may welcome the short term benefits of low energy prices, it seems certain that they will be a deterrent against energy conservation and will stop the oil and gas industry from seeking new domestic reserves. Low prices will also deter the creation of an industry to turn coal into synthetic oil and gas, and a shale oil industry. The 'deregulation' issue was another political debate that gained added importance because of a coming Presidential election (in 1976).

Reserve reservations

The post-crisis orgy of energy soul-searching lacked some of the reality that is essential to a meaningful understanding of the situation. People started talking about the world's energy reserves far more than they had before 1973, and unfortunately some of the newcomers to energy punditry failed to understand the status of energy reserves figures. Reserves estimates such as those prepared for the World Energy Conference (see chapter 2) were accorded a status that far exceeded their merit. There is nothing basically wrong with the WEC estimates or those of other organisations; but, as their compilers admit, they are far from perfect. For example, the massive numbers for the world's coal reserves (see table 2.2) give little indication of the recoverability of that coal.

In his book *Availability of world energy resources*, Dan Ion, who was an expert on energy reserves long before it became a fashionable topic, warned that our knowledge of the world's energy resources is poor. And a lot of work remains to be done to evaluate the existing estimates of the world's energy resources before we can look upon them as reliable enough inputs to decision making. Ion says that the first priority is to improve our knowledge of the world's *proved* reserves, rather than the more speculative potential reserves. This must be an early energy R&D priority, as must the development of new technologies to improve our knowledge of the whereabouts and extent of energy resources.

Fundamental questions

Other critics of energy orthodoxy have questioned some of society's cherished notions. One long held idea is that there is an unbreakable relationship between energy consumption and economic activity. As national wealth has risen – often measured in terms of the GNP – so has energy consumption. Various analysts of the statistics reckon they can detect a direct relationship between GNP and energy consumption.

The energy crisis hit the world just about the same time as the 'zero growth' movement got under way. Reports such as the famous *Limits to growth* questioned the belief that individuals could look forward to an ever rising standard of living based on increased consumption of raw materials, energy, water, food and the other commodities that make life comfortable in the developed nations. The shallowness and obvious mistakes behind much of the 'beware the end is nigh' school of thought did not stop the anti-growth idea from making a considerable impact. The debate sparked off by *The limits to growth* included discussion of the relationship between GNP and energy consumption. The question is: 'Will energy consumption continue to rise along with GNP?'

Bent Sørensen, Associate Professor of Physics at the Niels Bohr Institute, University of Copenhagen, Denmark, asked a more subtle question. He asked how much of the rise in energy consumption that had gone into increasing economic activity actually improved the standard of living of the people. Sørensen pointed out that while there may be a direct relationship between GNP and energy consumption, some countries – most notably Japan and Europe – had used half as much energy as the US in pushing up their GNP by one unit. He also pointed out that GNP can be a poor measure of the standard of living – economic activity includes activities that may have little or no social benefit. Sørensen maintains that as GNP rises, less and less of this increase finds its way back to consumers as a rise in the standard of living. Much of the increased activity can go into the structural changes needed to sustain growth, and to cope with its side effects – such as managing the increasing waste problems that go with growth.

If Sørensen is right, then perhaps we should organise the energy system in such a way that we do not merely increase the amount of energy available, but increase the direct benefits to the consumer. This might mean concentrating on renewable energy sources such as solar energy, wind power, and the like, rather than nuclear energy or some other high technology requiring an expensive and large industrial infrastructure. It might prove more expensive to produce a unit of energy by this 'low technology' approach, but the direct benefits to the consumer might cost less than energy from high technology systems. Sørensen was writing about the energy situation in Denmark – he reckons that wind power is already competitive with nuclear power in Denmark – but his comments could be true for other countries.

Sørensen is not an extremist in the cause of alternative energy systems. He acknowledges that the world's energy future could follow one of two paths. If fusion power, for example, can be made to work, the world might have unlimited supplies of inexpensive energy supplied by huge power stations. On the other hand, the price of consumable energy supplies might rise as they run out, and if nothing is there to take their place we will have to rely on renewable energy sources. The path we are forced along will clearly have a dramatic impact on the shape of society. An unlimited supply of inexpensive power from fusion reactors, for example, would make it easier to recycle waste and turn it into useful materials. Recycling of waste materials – an increasingly fashionable subject – will not be possible without considerable amounts of energy. Such a development – plenty of energy to turn garbage back into something useful – would relax the 'limits to growth'. The opponents of our polluting society should also acknowledge that anti-pollution measures consume energy. Hence 'clean' energy could be the key to a less rapacious and an environmentally more benign society.

Some advocates of alternative technology take an extreme line – they want nothing to do with high technology – and want to see energy production, and just about everything else, decentralised. (They dismiss the idea of a high energy, low pollution society, even though it would do away with many of the ills they dislike in society.) In the world of the alternative technology (AT) advocate, power would come from small units – homes would be 'autonomous', that is they would meet most of their own needs. Each house would have a collection of energy systems including windmills, solar energy systems, heat pumps, and so on. While this approach may not be appropriate to the needs of most home occupiers – few of the AT experiments seem to concern themselves with the average dwelling of the sort occupied by most of the population – energy systems based on renewable resources could play a part in the average home. The problem is to find the right balance, and to employ the right technology where it is appropriate.

The future shape of the world's energy system is anybody's guess – there are as many alternative projections of energy supply and demand as there are energy pundits. And each projection can have several estimates of how different fuels will contribute to the future energy scenario. It is impossible to make really accurate forecasts that take in all of the variables – and no forecast can cope with the vagaries of human nature. All the sensible forecaster can do is produce broad bands within which the future energy curves will *probably* fall. Most forecasts confidently predict that energy consumption will continue its upward growth as economic growth continues; but few observers of the energy scene expect a return to the energy profligacy of the 1950s and 1960s. Even before the 1973/4 oil crisis more perceptive observers of the energy business were warning that the situation would change for the worse. Unfortunately, few of the people in a position to react to these warnings took much notice. The events of 1973/4 soon changed that, but the lessons of the past few years may well be forgotten as they become history. When energy recedes from the headlines, politicians may find it very tempting to save money by taking apart the expensive energy R&D programmes they are now busily assembling.

One lesson we should learn from recent events is that flexibility is a desirable commodity in the energy business. This means not placing too much reliance on oil, as was the case in the past; or on nuclear power, as could be the situation in the future, given the disproportionate share of energy R&D funds that now goes to nuclear research. Even supporters of nuclear power fear that the first serious reactor accident could bring the business to at least a temporary halt – and it takes some conviction to believe that there will never be a serious nuclear accident. (The need for an 'insurance policy' against such an accident was one of the reasons given for Britain's creation of a wave energy R&D capability.)

Given that diversity will be required of the world's energy system, it seems inevitable that many, if not most, of the different energy technologies described in earlier chapters will have some part to play – just how big this part will be will depend on events. It seems likely, however, that we will pay more attention to locally exploitable resources. Wind energy, for example, will be used where it is appropriate. It will not be ignored simply because it will be useless (and there will be no market for wind power hardware) in many parts of the world. There are signs that wind power – which is perhaps the alternative energy system nearest to commercial viability – already is economic in remote and windy parts of Britain and other countries.

One way in which diversity should come about is through the use of the fuel and energy technology that is most appropriate for the job in hand. Thus space heating would, where the climate is amenable, be the province of solar energy, which could also meet domestic hot water requirements even in regions where the climate is not good enough to allow solar space heating. Given that only so much money can be devoted to solar energy R&D, it makes less sense to spend money on projects that would lead to massive solar power stations where there is already plenty of competition from nuclear power, for example.

If there is to be this diversity of energy supply, it will have to come because people make the right decisions. And this means that the energy consumer has to be more aware of the energy situation and the impact of his actions. It is tempting to leave everything to the electricity company, or the fuel suppliers: you need very little knowledge to pick your way through such a simple system. The added complexity of a variety of renewable energy technologies will confuse things no end.

Clearly if we want to stretch the world's reserves of fossil fuels, and particularly the supplies of natural gas and oil, we should not use them where there are perfectly adequate alternatives based on more plentiful energy supplies. This means that the near criminal combustion of natural gas in power stations has to stop. This is no longer a widespread practice, but the world's gas reserves are so small as to allow no such misuse. Even the combustion of oil in power stations is something we should think very seriously about. I do not subscribe to the sentiment that oil is too precious to burn – there simply is not the need for petrochemical feedstocks to warrant a burn-no-oil policy – but it does not do to burn oil simply because it is there. Oil companies and electricity utilities tell us that power station fuel oil is a 'waste' product left over after the desirable fractions of crude oil have been extracted; but if they really wanted to, the petrochemical engineers could turn fuel oil into more valuable commodities, just as they now talk of turning coal into liquid fuels.

Wind power, geothermal energy, tidal power, and all the other bit players in the energy picture will find suitable roles. Just because geothermal energy, say, will be able to supply only a few per cent of Britain's energy does not mean that we should ignore it. Other nations operate energy systems smaller than the contribution the UK might expect geothermal energy to make to its energy supplies.

Of course, a complex energy system of the sort I advocate will not be the easiest thing to manage. And the record of the world's energy managers shows that most nations have mismanaged their simple three- or four-fuel economies. I suspect that many of today's energy R & D portfolios are being assembled on the assumption that one topic will pay off so handsomely that eventually everything else will be dropped, leaving a single energy technology that will meet all needs into the foreseeable future. (This was certainly the motivation behind earlier nuclear R & D programmes, and would probably continue to be so were it not for the opprobrium surrounding nuclear energy.) Life might turn out as simple as this, but we would be better off working under the assumption that things will be more complicated in the future. If nuclear fission, or fusion, or even solar energy proves to be the most versatile of energy sources with no limits on its applicability, then we can adapt to meet the situation. It will be infinitely harder if we put all our eggs in one basket, only to discover that there is a hole in the bottom. On the other hand, it will not do to scatter R & D funds so widely that no energy technology receives enough support to prove itself and make the difficult transitions from research through to development, and finally on to production and proliferation. Clearly the managers of the world's energy R & D programmes have an onerous task: if they fail to perform it skilfully the future energy picture could look decidedly gloomy.

Information

There is no central source of information on energy research and development. There is little point in giving here a long list of scientific papers, no matter how good they are, when the would-be reader has little chance of getting to them. Fortunately, someone has already gathered together a first class anthology of papers by all the right people on all the relevant subjects:

Perspectives on energy, edited by Lon C. Ruedisili and Morris W. Firebaugh, Oxford University Press. This anthology was published too late to help me write this account of energy R & D, but it includes several of the papers I referred to in my research, and also other contributions by authors whose work has provided information for this book. The papers are almost all American; but the US has, after all, done most to further the study of energy technology.

Energy statistics are reasonably accessible, although it isn't always easy to assess their accuracy. The United Nations regular series of energy statistics is sometimes a little out of date, but it is the most comprehensive compilation of energy consumption data available. Most governments also publish regular energy statistics. Three of the most useful publications are:

Survey of energy resources 1974, World Energy Conference, September 1974. This report, which anyone can buy, provided much of the information for chapter 2. It is probably the most thorough compilation of statistics on world energy reserves and resources. There have been arguments about its accuracy, but it is the best study available.

BP statistical review of the world oil industry, British Petroleum Company, London. Each year BP produces this invaluable statistical survey. It is a much-quoted source of useful numbers.

Uranium: resources, production, and demand, IAEA/OECD. The OECD's Nuclear Energy Agency in Paris is a good source of information on uranium resources and radioactive waste management; the NEA also supervises the dumping at sea of radioactive waste. This survey is produced every couple of years or so. The latest edition – which came out too late for this book – was published early in 1976.

General reports on energy research and development are available from numerous organisations. Most countries, for example, with the possible exception of the UK, have published a report on their energy R & D prospects. The list starts with an international report:

Energy R & D, OECD, 1975. A good, if uncritical survey of energy technologies and R & D policies in OECD member countries. It was used in the preparation of the report *Energy prospects to 1985* (see later in this list).

The nation's energy future (WASH-1281), US Atomic Energy Commission, December 1973. President Nixon asked for this report in June 1973 – before the world's energy supplies were so rudely disrupted. The report was put together in a great hurry and is a heavily condensed version of a mountain of study group papers prepared for WASH-1281. Dixy Lee Ray, chairman of the USAEC at the time, oversaw the production of this report, which had quite an influence when it appeared.

A national plan for energy research, development and demonstration (ERDA-48), US Energy Research and Development Administration, June 1975. ERDA's 'blueprint' for energy R & D in the US was also put together in a hurry; however, it remains a good follow-up to WASH-1281, with a promise that it will be regularly updated.

An analysis of the ERDA plan and program, 1975, by the Office of Technology Assessment – an office set up to keep the US Congress informed about technological issues and the possible impact of the government's technological actions on society. This critical assessment of ERDA's efforts shows that in the US at least the energy R & D establishment will not be able to get away with a programme based on tired old notions and strictly conventional thinking.

US energy prospects: an engineering viewpoint, National Academy of Engineering, May 1974. This report is particularly interesting because the man who was president of the NAE at the time and a leading light in the study, Robert Seamans, later became head of ERDA.

Energy conservation: a study by the Central Policy Review Staff, HMSO, 1974. The CPRS, the 'think tank' of the British Cabinet Office, threw the cat among the pigeons when it was forced, against its will, to issue this report. It is not, despite its title, a report on energy conservaton in the strict sense. It also talks about energy supply technologies. This report has the distinction of being the first UK government report to suggest that the country should look closely at wave power.

Peaceful uses of atomic energy, UN/IAEA. A 15-volume library on all aspects of nuclear energy. The Fourth International Conference on the Peaceful Uses of Atomic Energy was held in 1971, but these proceedings of the conference remain a good source of information on atomic energy.

The nuclear industry 1974 (WASH-1174-74). The USAEC published an 'annual report' on the current status of the US nuclear industry each year.

Solar energy as a national energy source, US National Science Foundation/National Aeronautics and Space Administration, December 1972. This was one of the earlier official reports to take solar energy seriously.

Solar energy for earth, American Institute for Aeronautics and Astronautics, April 1975. A good review of solar energy technology and its prospects.

Geothermal energy: resources, production, stimulation, edited by Paul Kruger and Carel Otto, Stanford University Press, 1973. The best book

I have seen on geothermal energy. This is an anthology, covering most aspects of the subject.

The increased cost of energy – implications for UK industry, National Economic Development Office, June 1974, and *Energy conservation in the United Kingdom*, also NEDO, January 1975. These two reports show why we should worry about the future of the world's energy supplies – dearer energy pushes up all costs.

Energy conservation: ways and means, The Future Shape of Technology Foundation, The Hague, 1974. An early contribution on how we might use energy less wastefully. Work on this report started in 1971, which may explain why its thoroughness and quality have been unequalled by most of the subsequent reports that have been assembled on the same subject.

Energy prospects to 1985, OECD, January 1975. 'An assessment of long term energy developments and related policies.' This two-volume report grabbed the headlines when it was published. It outlines some of the factors that go into the formulation of national energy policies. Its R&D input came from *Energy R & D* (see above).

A time to choose: America's energy future, The Energy Policy Project of the Ford Foundation, October 1974, Ballinger Publishing Company, Cambridge, Massachusetts. The $4 million Energy Policy Project produced some 20 reports on the energy situation, with a strong American emphasis. This final report was extremely controversial, provoking one oil company to buy advertising space in the *New York Times* to refute the report's conclusions. There was also at least one 'anti-report'.

No time to confuse, Institute for Contemporary Studies. This collection of critical essays – from such eminent thinkers as Herman Kahn and Morris Adelman – tore into the above report. It criticised the Energy Policy Project's economic naivety, and questioned the motives of the project's staff.

Acknowledgements

Illustrations are reproduced by permission of the following: p. 52: National Coal Board; pp. 56, 172, 202: Central Electricity Generating Board; pp. 74, 80: British Petroleum; pp. 100, 126: U.K Atomic Energy Authority; p. 106: British Nuclear Fuels Ltd; p. 140: Honeywell; p. 144: Centre National de la Recherche Scientifique; pp. 156: Ente Nazionale per l'Energia Elettrica; p. 168: Electricité de France; pp. 178: NASA; p. 194: Novosti Press Agency.

Cartoons are reproduced by kind permission of Richard Willson.

Index